Understanding Evolution

Fifth Edition

Understanding Evolution

Fifth Edition

E. Peter Volpe
Mercer University

wcb

Wm. C. Brown Publishers
Dubuque, Iowa

*To my
ever-probing students
who taught me
how to teach*

Contents

Preface

The publication of Charles Darwin's *The Origin of Species* in 1859 is one of the great landmarks in the history of science. Darwin brought about one of the most remarkable and far-reaching revolutions in human thought. Before Darwin, we arrogantly presumed that we occupied an exalted position in the world. Darwin removed us from the center stage of the earth and made us part of the fabric of nature. Although Darwin's monumental book was greeted at first with a storm of criticism, the sheer weight of evidence inevitably forced conviction of the truth of evolution even upon unwilling minds.

In recent years, another provocative book has appeared on the scene. Edward O. Wilson's *Sociobiology: The New Synthesis,* published by the Harvard University Press in 1975, has received wide attention but not universal recognition. The voluminous *Sociobiology* has been the subject of controversy, more or less acrimonious, from the time of its first appearance to the present day. The Harvard biologist Edward Wilson has proposed to explain human actions—particularly social behavior—on the basis of evolutionary concepts. Understandably, the notion that human sociality is influenced by natural evolutionary forces is alien to the social sciences and humanities, where the outlook of human behavior is largely devoid of reliance on biological principles. Wilson's insistence on a biological underpinning to the social interplay of humans is essentially an extension of Darwin's tenets. The new sociobiology relies on the evolutionary processes for interpretations and predictions. Accordingly, an appreciation of the views of sociobiology requires a keen understanding of Darwinian evolution. For this reason, the section on sociobiology in this text is sensibly reserved as the concluding chapter.

Darwin portrayed evolution as a slow process, proceeding by fine, graduated steps. In recent years, Stephen Jay Gould of Harvard University and Niles Eldredge of the American Museum of Natural History have asserted that the fossil record does not support the notion of gradual change. Rather, they contend that new forms of life entered the fossil record abruptly and then persisted with little significant change until extinction. The presumed pattern of the sudden appearance of new species followed by very long periods of unexpected stability has been termed *punctuated equilibria* or, colloquially, "evolution by leaps and bounds." This new school of thought has attracted critical attention, and is the subject of a new chapter (17) in this revised text.

This updated edition of *Understanding Evolution* is a blend of the time-honored, established body of information and the newer knowledge. Topics of current interest that command attention in this revised text are sociobiology, origin of life, punctuated equilibria, molecular evolution, and the evolutionary status of humans. Notwithstanding the readjustment of content between established and new, the prime goal of this edition, like the earlier ones, remains the same. I have endeavored to present a simple, concise account of the scope and significance of evolution for the college student seeking a liberal education. I hope that the account may dispel the vague and naive notions about organic evolution held by many beginning students. They view evolution as something that happened in the remote past. They know it has something to do with dinosaurs, rocks, and the proverbial "missing link." And they are familiar with alluring clichés as "struggle for existence" and "survival of the fittest." But few pause to consider how the process of evolution actually works. In the belief that any concept can be best understood by knowing how it operates, I have focused on the mechanism of evolution, the causal aspect rather than the historical. Particular attention has been given to modern observations and experiments that illustrate and clarify the evolutionary process. Several examples of changes in human populations have been cited so that the student may realize that evolution in mankind has not come to a standstill.

Evolution is a process of continual change. Organisms throughout life's history have not remained constant but have gradually and endlessly changed. Change is the rule of living things. The occurrence of organic evolution does not in itself reveal *how* evolution is brought about. That is, we may know of an event or phenomenon and accept it as true, even though we may not fully understand the forces that determine its existence. Scientists no longer debate the reality of evolution. However, differences of opinion have arisen in the *explanation* of evolution. One may challenge an interpretation, but to contest the interpretation is not to deny the existence of the event itself. A widespread fallacy is to discredit the truth of evolution by seizing on points of disagreement concerning the mechanism of evolution.

Charles Darwin was the first person to reach a meaningful understanding of evolution. The concept of natural selection that he masterfully put forward in *The Origin of Species* remains the keystone of the evolutionary process. The Darwinian principle of natural selection is now firmly established as the main driving force of evolution. However, the Darwinian thesis has been enriched and refined by recent advances in systematics, ecology, cytology, paleontology, and, above all, in genetics. The modern extension of the great work of Gregor Mendel in heredity has profoundly influenced current thoughts about evolution. The mechanism of evolution is incomprehensible without familiarity with the fundamentals of genetics. Thus, the basic principles of genetics needed for an understanding of evolution are included in this text.

The text is clarified by numerous illustrations. The drawings, many of them original, are the accomplished work of my wife, Carolyn. The illustrations do not merely adorn the pages; all are important in supporting the writing. Selected references accompany each chapter in the hope that the student will want to extend his knowledge beyond the material presented.

The book could not have been prepared without the knowledge derived from my past teachers and students. The latter, perhaps unknowingly, have also been my teachers. If simplicity and clarity of presentation of the subject matter have been achieved in this book, it will be due in great measure to the long-suffering college freshmen in my introductory general biology course over the past twenty years or so. I appreciate the helpful suggestions and advice throughout the years from numerous colleagues, who hopefully will not be offended if I fail to list their names. They know, of course, that I am grateful. The presentation has benefited from the editorial work of Wilma Parrish, who was a perceptive critic in every phase of the revision. Finally, I am indebted to the authors and publishers who have generously granted permission to use figures and tables from their books. Individual acknowledgments are made where the figures and charts appear in the text.

E. Peter Volpe
Macon, Georgia

Understanding
Evolution

1
Meaning of Evolution

In the fall of 1958, the folks of a quiet, rural community in the southern part of the United States were startled and dismayed by the large numbers of *multilegged* bullfrogs in an 85-acre artificial lake on a cotton farm. Widespread newspaper publicity of this strange event attracted the attention of university scientists, curiosity seekers, and gourmets. The lake supports a large population of bullfrogs, estimated at several thousand. Although reports tended to be exaggerated, there were undoubtedly in excess of 350 multilegged deviants. As illustrated in figure 1.1, the extra legs were oddly positioned, but they were unmistakably copies of the two normal hind limbs. Incredibly the extra limbs were functional, but their movements were perceptibly not in harmony with the pair of normal legs. The bizarre multilegged frogs were clumsy and graceless.

All the multilegged frogs appeared to be of the same age, approximately two years old, and of the same generation. These atypic frogs were found only during the one season, and were not detected again in subsequent years. The multilegged frogs disappeared almost as dramatically as they had appeared.

Strange and exceptional events of this kind are of absorbing interest, and challenge us for an explanation. How do such oddities arise and what are the factors responsible for their ultimate disappearance in a natural population? In ancient times, bodily deformities evoked reverential awe and inspired some fanciful tales. Early humans constructed a number of myths to explain odd events totally beyond their control or comprehension. One legend has it that when masses of skeletons are revived or reanimated, the bones of different animals often become confused. Another old idea is that grotesquely shaped frogs are throwbacks to some remote prehistoric ancestor. These accounts are, of course, novelistic and illusory. They do, however, reveal the uniquely imaginative capacity of the human mind. Nevertheless, we should seek a completely different cause-and-effect sequence, relying more on our faculty for logical analysis.

Figure 1.1 **Two multilegged frogs,** each viewed from the back (dorsal) and front (ventral) surface. These bizarre bullfrogs were discovered in October, 1958, in a lake near Tunica, Mississippi. Several hundred frogs with extra hind limbs were found at this locality. How does such an abnormality arise? Two reasonable interpretations are set forth in chapter 1.

MULTILEGGED BULLFROGS

Environmental Modification

Inspection alone cannot reveal the underlying cause of the multilegged anomaly. We may thoroughly dissect the limbs and describe in detail the anatomy of each component, but no amount of dissection can tell us how the malformation arose. The deformity either was foreordained by heredity or originated from injury to the embryo at a vulnerable stage in its development. We shall direct our attention first to the latter possibility and its implications.

Some external factor in the environment may have adversely affected the pattern of development of the hind limb region. The cotton fields around the pond were periodically sprayed with pesticides to combat noxious insects. It is not inconceivable that the chemicals used were potent *teratogens* —that is, substances capable of causing marked distortions of normal body parts.

That chemical substances can have detrimental effects on the developing organism is well documented. For example, the geneticist Walter Landauer demonstrated in the 1950s that a wide variety of chemicals, such as boric acid, pilocarpine, and insulin (normally a beneficial hormone), can produce abnormalities of the legs and beaks when injected into chick embryos. This type of finding cannot be dismissed complacently as an instance of a laboratory demonstration without parallel in real life. Indeed, in 1961, medical researchers discovered with amazement amounting to incredulity that a purportedly harmless sleeping pill made of the drug thalidomide, when taken by a pregnant woman, particularly during the second month of pregnancy, could lead to a grotesque deformity in the newborn baby, a rare condition in humans called phocomelia—literally, "seal limbs." The arms are absent or reduced to tiny, flipperlike stumps (fig. 1.2).

In different organisms—fish, frogs, mice, and humans—anomalies can be caused by such diverse agents as extremes of temperature, X rays, viruses, drugs, diet deficiencies, and lack of oxygen. More than a century ago, in 1832,

---

Figure 1.2 Armless deformity in the newborn infant, resulting from the action of thalidomide, a sedative taken by his mother in her second month of pregnancy. (From W. Lenz and K. Knapp, "Thalidomide Embryopathy," *Archives of Environmental Health* 5 (1962):100–105. Courtesy of the publisher and Dr. W. Lenz.)

the French biologist Etienne Geoffroy St. Hilaire sealed the air pores in the shell of a hen's egg with varnish, and observed that the embryo, deprived of its oxygen supply, became deformed. Experiments by modern investigators, like Theodore Ingalls of Harvard University, have confirmed St. Hilaire's crude but informative experiment. Deprivation of oxygen during the early divisions of fertilized eggs of the zebra fish can lead to the formation of abnormally small eyes or a single eye only. In 1941, the medical community was startled by the observation that pregnant women who had contracted German measles during the first three months of pregnancy often gave birth to blinded infants. The infants were afflicted with cataract, a condition in which the eye lens is opaque, obstructing the passage of light. There is no longer the slightest doubt that environmental factors may be causal agents of specific defects. A variation that arises as a direct response to some external change in the environment, and not by any change in the genetic makeup of the individual, is called an *environmental modification.*

If the multilegged anomaly in the bullfrog was environmentally induced, we may surmise (as depicted in figure 1.3) that the harmful chemical or other causative factor acted during the sensitive early embryonic stage. Moreover, if some external agent had brought about the abnormality in the bullfrog, this agent must have been effective only once, because the multilegged condition did not occur repeatedly over the years. It may be that the harmful environmental factor did not recur, in which event we would not expect a reappearance of the malformation.

Our suppositions could be put to a test by controlled breeding experiments, the importance of which cannot be overstated. Only through breeding tests can the basis of the variation be firmly established. If the anomaly constituted an environmental modification, then a cross of two multilegged bullfrogs would yield all normal progeny, as illustrated in figure 1.3. In the absence of any disturbing environmental factors, the offspring would develop normal hind limbs. This breeding experiment was not actually performed; none of the malformed frogs survived to sexual maturity. Nonetheless, we have brought into focus an important biological principle: *environmentally induced (nongenetic) traits cannot be passed on to another generation.*

The elements that are transmitted to the next generation are two minute cells, the egg and the sperm. These two specialized cells are often referred to as *germ* cells, because they are the beginnings, or germs, of new individuals. The germ cells represent the only connecting thread between successive generations. Accordingly, the mechanism of hereditary transmission must operate across this slender connecting bridge. The hereditary qualities of the offspring are established at the time the sperm unites with the egg. The basic hereditary determiners, the *genes,* occur in pairs in the fertilized egg. Each inherited characteristic is governed by at least one pair of genes, with one member of each pair coming from the male parent and the other from the female parent. Stated another way, the sperm and egg each has half the number of genes of the parents. Fertilization restores the original number and the paired condition (fig. 1.3).

Figure 1.3   **Test for the noninheritance of an environmentally induced trait.** If the malformation of the hind limbs arose from without by the action of an adverse environmental agent (pesticide) and not from within by genetic change, then the abnormality would *not* be transmitted from parent to offspring. The multilegged parents, although visibly abnormal, are both genetically normal. Normal hind limbs are expected if, during development, the offspring are not exposed to the same injurious environmental agents as were the parents.

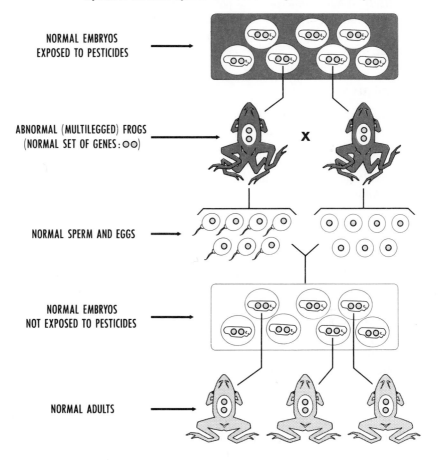

NORMAL EMBRYOS
EXPOSED TO PESTICIDES

ABNORMAL (MULTILEGGED) FROGS
(NORMAL SET OF GENES: ○○)

NORMAL SPERM AND EGGS

NORMAL EMBRYOS
NOT EXPOSED TO PESTICIDES

NORMAL ADULTS

The underlying assumption in the cross depicted in figure 1.3 is that the genes that influence the development of the hind limbs are normal ones. Of course, the multilegged parents themselves possess normal genes. It might seem strange that an abnormal character can result from a perfectly sound set of genes. However, normal genes cannot be expected to act normally under all environmental circumstances. A gene may be likened to a photographic negative. A perfect negative (normal gene) may produce an excellent or poor positive print (normal or abnormal trait), depending on such factors (environmental factors) as the quality or concentration of the chemical solutions used in preparing the print. The environment thus affects the *expression* of the negative (gene), but the negative (gene) itself remains unaffected throughout the making of the print (trait).

We shall learn more about the nature of genes and the mechanism of inheritance in the next chapter. For the moment, the important consideration is that *a given set of genes prescribes a potentiality for a trait and not the trait itself.* What is inherited is a potential capacity. The potential capacity will not become a developed capacity unless the appropriate environment is furnished. Genes always act within the conditioning framework of the environment.

Lamarckism

Few people would expect bodily deformities caused by harmful environmental factors to be inheritable. And yet, many persons believe that favorable or beneficial bodily changes acquired or developed during one's lifetime are transmitted to the offspring. As a familiar example, an athlete who exercised and developed large muscles would pass down his powerful muscular development to his children. This is the famous theory of the *inheritance of acquired characteristics,* or *Lamarckism* (after Jean Baptiste de Lamarck, a French naturalist of the late 1700s and early 1800s). The concept of Lamarckism has no foundation of factual evidence. We know, for example, that a woman who has altered her body by injections of silicone does not automatically pass the alterations on to her daughter. Circumcision is still a requisite in the newborn male, despite a rite that has been practiced for well over 4,000 years. It is sufficient to state that the results of countless laboratory experiments testing the possibility of the inheritance of acquired, or environmentally induced, bodily traits have been negative.

The notion of the inheritance of acquired characteristics became the cornerstone of Lamarck's comprehensive, although incorrect, explanation of evolution. Lamarck's theory is exemplified by a notable quotation from his book *Philosophie Zoologique* (1809):

> The giraffe lives in places where the ground is almost invariably parched and without grass. Obliged to browse upon trees it is continually forced to stretch upwards. This habit maintained over long periods of time by every individual of the race has resulted in the forelimbs becoming longer than the hind ones, and the neck so elongated that a giraffe can raise his head to a height of eighteen feet without taking his forelimbs off the ground.

Lamarck advocated that the organs of an animal became modified in appropriate fashion in direct response to a changing environment. The various organs became greatly improved through use or reduced to vestiges through disuse. Such bodily modifications, in some manner, could be transferred and impressed on the germ cells to affect future generations. Thus, the whale lost its hind limbs as the consequence of the inherited effects of disuse, and the wading bird developed its long legs through generations of sustained stretching to keep the body above the water level. Inheritance was viewed simply as the direct transmission of those superficial bodily changes that arose within the lifetime of the individual owing to use or disuse. Lamarck's views are unacceptable because we know of no mechanism that allows changes in the body to register themselves on the germ cells.

The foregoing considerations bear importantly on our bullfrog population. We have not been able to perform the critical breeding test to prove or disprove that the multilegged trait is a noninheritable modification. However, let us for the moment accept the thesis that the multilegged condition was environmentally induced and examine the implications of this view on the population as a whole. Outwardly, a striking change in the bullfrog population seems to have occurred. But in actuality the composition of the population has remained essentially unaltered, since the basic hereditary materials, the genes, were not affected. In other words, there was no change of evolutionary significance in the population. Evolution can occur *only* where there is inheritable variation.

Inheritable Variation

We shall now examine the possibility that the multilegged frogs were genetically abnormal. One or more of their genes may have been impaired. Unlike teratogens, which damage an already conceived offspring and affect only a single generation, deleterious genes are in the germ plasm before conception and may lie dormant for several generations.

We shall postulate that the faulty gene for the multilegged condition is *recessive* to the normal gene. That is to say, the expression of the faulty gene is completely masked or suppressed by the normal (or *dominant*) gene when the two are present together in the same individual (fig. 1.4). Such an individual, normal in appearance but harboring the harmful recessive gene, is said to be a *carrier*. The recessive gene may be transmitted without any outward manifestation for several generations, continually being sheltered by its dominant partner. However, as seen in figure 1.4, the deleterious recessive gene ultimately becomes exposed when two carrier parents happen to mate. Those progeny that are endowed with two recessive genes, one from each parent, are malformed. On the average, one-fourth of the offspring will be multilegged.

Genetic defects transmitted by recessive genes are not at all unusual. Pertinent to the present discussion is an inherited syndrome of abnormalities in humans, known as the Ellis-van Creveld syndrome. Afflicted individuals are disproportionately dwarfed (short-limbed), have malformed hearts, and possess six fingers on each hand (fig. 1.5). The recessive gene that is responsible for this complex of defects is exceedingly rare. Yet, as shown by the geneticist Victor A. McKusick of Johns Hopkins University, the Ellis-van Creveld anomaly occurs with an exceptionally high incidence among the Amish people in Lancaster County, Pennsylvania. The defective recessive gene apparently was present in one member of the original Old Order Amish immigrants from

Figure 1.4   Mating of two "carrier" parents resulting in the emergence of multilegged offspring. The multilegged trait is assumed to be controlled by an abnormal recessive gene. Both parents are normal in appearance, but each carries the abnormal gene masked by the normal gene. The recessive gene manifests its detrimental effect when the offspring inherits one abnormal gene from each of its parents.

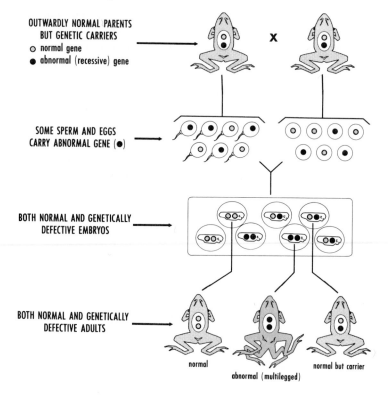

OUTWARDLY NORMAL PARENTS
BUT GENETIC CARRIERS
○ normal gene
● abnormal (recessive) gene

SOME SPERM AND EGGS
CARRY ABNORMAL GENE (●)

BOTH NORMAL AND GENETICALLY
DEFECTIVE EMBRYOS

BOTH NORMAL AND GENETICALLY
DEFECTIVE ADULTS

normal   normal but carrier
abnormal (multilegged)

Europe two centuries ago. For a few generations, the harmful gene was passed down unobserved, masked by its normal partner gene. Since 1860, the Ellis-van Creveld deformity has appeared in at least 50 offspring. Ordinarily, it is uncommon for both members of a married couple to harbor the abnormal recessive gene. However, in the sober religious Amish community, marriages have been largely confined within members of the sect with a resulting high degree of consanguinity. Marriages of close relatives have tended to promote the meeting of two normal, but carrier, parents.

We know very little about the past history or breeding structure of the particular bullfrog population in which the multilegged trait appeared with an exceptionally high frequency. We may suspect that all 350 or more multilegged frogs were derived from a single mating of two carrier parents. In contrast to humans, a single mated pair of frogs can produce well over 10,000

Figure 1.5 **Six-digited hands,** one of the manifestations of the Ellis-van Creveld syndrome in humans, a rare recessive deformity. Affected offspring generally come from two normal, but carrier, parents, each of which harbors the abnormal recessive gene. (Courtesy of Dr. Victor A. McKusick.)

offspring. The similarity in age of the multilegged frogs found in nature adds weight to the supposition that these frogs are members of one generation, and probably of one mating.

Much of the preceding discussion on the bullfrog is admittedly speculative. However, one aspect is certain: *previously concealed harmful genes are brought to light through the mechanism of heredity.* A trait absent for many generations can suddenly appear without warning. Once a variant character expresses itself, its fate will be determined by the ability of the individual displaying the trait to survive and reproduce in its given environment.

It is difficult to imagine that the grotesquely shaped frogs could compete successfully with their normal kin. However, we shall never know whether or not the multilegged frogs were capable of contending with the severities of climatic or seasonal changes, or of successfully escaping their predators, or even of actively defending themselves. What we do know is that the multilegged frogs did not survive to reproductive age. Despite diligent searches by many interested investigators, no sexually mature abnormal frogs have been uncovered in the natural population. It seems that this unfavorable variant has been eliminated. Let us now explore this situation in light of Darwin's theory of natural selection.

Darwinian Evolution

Slightly more than a century ago, in 1859, Charles Robert Darwin (fig. 1.6) gave the biological world the master key that unlocked all previous perplexities about evolution. His thesis of natural selection can be compared only with such revolutionary ideas as Newton's law of gravitation and Einstein's theory of relativity. The concept of natural selection was set forth clearly and con-

Figure 1.6   Charles Darwin at the age of 31 (1840), four years after his famous voyage around the world as an unpaid naturalist aboard H.M.S. *Beagle*. (From a watercolor by George Richmond; American Museum of Natural History.)

vincingly by Darwin in his monumental treatise *The Origin of Species*. This epoch-making book was the fruition of more than twenty years of meticulous accumulation and analysis of facts.

In 1831, Charles Darwin, then twenty-two years old and fresh from Cambridge University, accepted the unpaid post of naturalist aboard H.M.S. *Beagle*, a ship commissioned by the British Admiralty for a surveying voyage around the world. Although Darwin was an indifferent student at Cambridge, he did show an exceptional interest in the natural sciences. He was an ardent collector of beetles and enjoyed bird-watching and hunting.

It took the *Beagle* nearly five years—from 1831 to 1836—to circle the globe (fig.1.7). When Darwin first embarked on the voyage, he did not dispute the dogma that every species of organism had come into being at the same time and had remained permanently unaltered. He shared the contemporary view that all organisms had been created about 4000 B.C.—more precisely, at 9:00 A.M. on Sunday, October 23, in 4004 B.C., according to the extraordinary pronouncement of Archbishop James Ussher in the seventeenth century. Darwin had in fact studied for the clergy at Cambridge University. But he was to make observations on the *Beagle's* voyage that he could not reconcile with accepted beliefs.

Darwin's quarters on the *Beagle* were cramped, and he took only a few books on board. One of them was the newly published first volume of Charles Lyell's *Principles of Geology*, a parting gift from his Cambridge mentor, John Henslow, professor of botany. Lyell challenged the prevailing belief that the earth had been created by a divine plan merely 6,000 years ago. On the contrary, the earth's age could be measured in hundreds of millions of years. Lyell argued that the earth's mountains, valleys, rivers, and coastlines were shaped not by Noah's Flood but by the ordinary action of the rains, the winds, earthquakes, volcanoes, and other natural forces. Darwin was impressed by Lyell's emphasis on the great antiquity of the earth's rocks, and gradually came to perceive that the characteristics of organisms, as well as the face of the earth, could change over a vast span of time.

The living and extinct organisms that Darwin observed in the flat plains of the Argentine pampas and the Galápagos Islands sowed the seeds of Darwin's views on evolution. From old river beds in the Argentine pampas, he dug up bony remains of extinct mammals. One fossil finding was the massive *Toxodon*, whose appearance was likened to a hornless rhinoceros or a hippopotamus (fig. 1.8). Another fossil remain that attracted Darwin's attention was the skeleton of *Macrauchenia*, which he erroneously thought was clearly related to the camel because of the structure of the bones of its long neck. Other remarkable creatures were the huge *Pyrotherium*, resembling an elephant, and the light and graceful single-toed *Thoatherium*, rivaling the horse. The presence of *Thoatherium* testified that a horse had been among the ancient inhabitants of the continent. It was the Spanish settlers who reintroduced the modern horse, *Equus*, to the continent of South America in the sixteenth century. Darwin marveled that a native horse should have lived and

Figure 1.7   Five-year world voyage of H.M.S. *Beagle*. Darwin's observations on this voyage convinced him of the reality of evolution. Particularly impressive to him were the fossil skeletons of mammals unearthed in the Argentine pampas (see fig. 1.8) and the variety of tortoises and birds in the small group of volcanic islands, the Galápagos Islands (see fig. 11.3). (American Museum of Natural History.)

Voyage of H.M.S. Beagle (1831 - 1836)

Figure 1.8 **Curious hoofed mammals** (ungulates) that flourished on the continent of South America some 60 to 70 million years ago, and have long since vanished from the scene. Bones of these great mammals were found by Darwin on the flat treeless plains of Argentina.

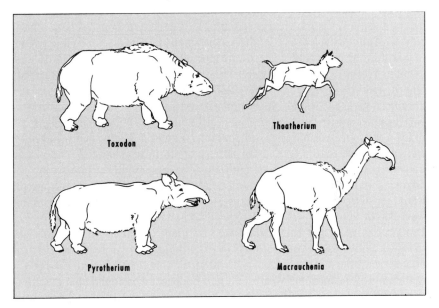

disappeared in South America. This was one of the first indications that species gradually became modified with time, and that not all species survived through the ages.

When Darwin collected the remains of giant armadillos and sloths on an Argentine pampa, he reflected on the fact that, although they clearly belonged to extinct forms, they were constructed on the same basic plan as the small living armadillos and sloths of the same region. This experience started him thinking of the fossil sequence of a given animal species through the ages and the causes of extinction. He wrote: "This wonderful relationship in the same continent between the dead and the living will, I do not doubt, hereafter throw more light on the appearance of organic beings on our earth, and their disappearance from it, than any other class of facts."

What Darwin had realized was that living species have ancestors. This fact is commonplace now, but it was a revelation then. On traveling from the north to the south of South America, Darwin observed that one species was replaced by similar, but slightly different, species. In the southern part of Argentina, Darwin caught a rare species of ostrich that was smaller and differently colored from the more northerly, common American ostrich, *Rhea americanus*. This rare species of bird was later named after him, *Rhea darwini*. It was scarcely imaginable to Darwin that several minor versions of a species would be created separately, one for each locality. It appeared to

Darwin that species change not only in time but also with geographical distance. He later wrote: "It was evident that such facts could only be explained on the supposition that species gradually became modified; and the subject haunted me."

In the Galápagos Islands, Darwin's scientific curiosity was sharply prodded by the many distinctive forms of life. The Galápagos consist of an isolated cluster of rugged volcanic islands in the eastern Pacific on the equator about 600 miles west of Ecuador. One of the most unusual animals is the giant land-dwelling tortoise, which may weigh as much as 550 pounds, grow to 6 feet in length, and attain an age of 200 to 250 years. The Spanish word for tortoise, *galápago,* gives the islands their name. Darwin noticed that the tortoises were clearly different from island to island, although the islands were only a few miles apart. In isolation, Darwin reasoned, each population had evolved its own distinctive features. Yet, all the island tortoises showed basic resemblances not only to each other but also to relatively large tortoises on the adjacent mainland of South America. All this revealed to Darwin that the island tortoises shared a common ancestor with the mainland forms. The same was true of a group of small black birds, known today as *Darwin's finches.* Darwin observed that the finches were different on the various islands, yet they were obviously closely related to each other. Darwin reasoned that the finches were derived from an ancestral stock that had emigrated from the mainland to the volcanic islands and had undergone profound changes under the different conditions of the individual islands. Apparently, a single ancestral group can give rise to several different varieties or species.

Mechanism of Evolution

When Darwin left England in 1831, he had accepted the established theory that different species had arisen all at once and remained unchanged and constant. His scientific colleagues did not deny the presence of fossil remains of species no longer existent. Such extinct species were thought to represent organisms that had been periodically annihilated by sudden and violent catastrophes. Each cataclysmic destruction was said to be accompanied by the creation of new forms of life, each distinct from the fossilized species. The theory of successive episodes of catastrophic destruction and creation, formulated by the French anatomist Georges Cuvier, was accepted by most eighteenth-century scholars.

On his return to England in October 1836, Darwin was convinced of the truth of the idea of descent—that all organisms, including humans, are modified descendants of previously existing forms of life. Darwin's observations during his voyage assured him that animals and plants are not the products of special creation, as Victorian society believed. We can discern two stages in the development of Darwin's thoughts: the first was the realization that the world of organisms is not fixed and unchangeable; the second was an explanation of the process of evolutionary change.

At home, Darwin assembled data relevant to the mechanism of evolution. He became intrigued by the extensive domestication of animals and plants brought about by conscious human efforts. Throughout the ages, humans have been a powerful agent in modifying wild species of animals and plants to suit their needs and whims. Through careful breeding programs, we determine which characteristics or qualities will be incorporated or discarded in our domesticated stocks. By conscious *selection,* we perfected the toylike Shetland pony, the Great Dane dog, the sleek Arabian race horse, and vast numbers of cultivated crops and ornamental plants.

Many clearly different domestic varieties have evolved from a single species through our efforts. As an example, we may consider the many varieties of domestic fowls (fig. 1.9), all of which are derived from a single wild species of red jungle fowl *(Gallus gallus)* present at one time in northern India. The red jungle fowl dates back to 2000 B.C. and has been so modified as to be nonexistent today. The variety of fowls perpetuated by fowl fanciers ranges from the flamboyant ceremonial cocks (the Japanese *Onaga-dori*) to the leghorns, bred especially to deposit spotless eggs.

Figure 1.9 Evolution under domestication. A variety of domestic fowls have evolved from a single wild species (now extinct) as a result of man's continual practice of artificial selection.

BLACK-TAILED JAPANESE BANTAM

FULL-SIZED AND BANTAM WHITE LEGHORN

SEBASTOPOL GOOSE

ONAGA-DORI COCK

CRESTED WHITE DUCK

FRIZZLED SULTAN ROOSTER

BIRCHEN GAME BANTAM

ARAUCANA

In the late 1830s, Darwin attended the meetings of animal breeders and intently read their publications. Animal breeders were conversant with the variability in their pet animals, and dwelled on the technique of *artificial selection.* That is, the breeders selected and perpetuated those variant types that interested them or seemed useful to them. The breeders, however, had only vague notions as to the origin, or inheritance, of the variable traits.

Darwin acknowledged the unlimited variability in organisms, but was never able to explain satisfactorily how a variant trait was inherited. He and other naturalists were unaware of Gregor Mendel's contemporaneous discovery. Rather, Darwin followed Lamarck in assuming that biological variation is chiefly conditioned by direct influences of the environment on the organism. He believed that the changes induced by the environment became inherited; that bodily changes in the parents leave an impression or mark on their germ cells, or gametes. Darwin thought that all organs of the body sent contributions to the germ cells in the form of particles, or *gemmules.* These gemmules, or minute delegates of bodily parts, were supposedly discharged into the bloodstream and became ultimately concentrated in the gametes. In the next generation, each gemmule reproduced its particular bodily component. The theory was called *pangenesis,* or "origin from all," since all parental bodily cells were supposed to take part in the formation of the new individual.

Having explained the origin of variation (although incorrectly), Darwin wondered how artificial selection (a term familiar then only to animal breeders) could be carried on in nature. There was no breeder in nature to pick and choose. In 1837, Darwin wrote: "How selection could be applied to organisms living in a state of nature remained a mystery to me." Slightly more than a year later, Darwin found the solution. In his autobiography, Darwin explained: "In October 1838, that is fifteen months after I had begun my systematic enquiry, I happened to read for amusement *Malthus on Population . . .* at once it struck me that under these circumstances favourable variations would tend to be preserved and unfavourable ones destroyed. The result of this would be the formation of new species." The circumstances mentioned by Darwin were those associated with the assertions of an English clergyman, Thomas Robert Malthus, that population is necessarily limited by the means of subsistence.

Darwin's Natural Selection

Malthus's writings provided the germ for Darwin's thesis of *natural selection.* In his famous *An Essay on the Principle of Population,* Malthus expressed the view that the reproductive capacity of mankind far exceeds the food supply available to nourish an expanding human population. Humans compete among themselves for the necessities of life. This unrelenting competition engenders vice, misery, war, and famine. It thus occurred to Darwin that competition exists among all living things. Darwin then envisioned that the "struggle for existence" might be the means by which the well-adapted individuals survive and the ill-adjusted are eliminated. Darwin was the first to realize that perpetual selection existed in nature in the form of *natural selection.* In natural selection, as contrasted to artificial selection, the animal breeder or horticulturist is replaced by the conditions of the environment that prevent the survival and reproduction of certain individuals. The process of natural selection occurs without a conscious plan or purpose. Natural selection was an entirely new concept, and Darwin was its proponent.

It was not until 1844 that Darwin developed his idea of natural selection in an essay, but not for publication. He showed the manuscript to the geologist Charles Lyell, who encouraged him to prepare a book. Darwin still took no steps toward publishing his views. It appears that Darwin might not have prepared his famous volume had not a fellow naturalist in the Dutch East Indies, Alfred Russel Wallace, independently conceived of the idea of natural selection. Wallace had spent many years exploring and collecting in South America and the East Indies. Wallace was also inspired by reading Malthus's essay, and the idea of natural selection came to him in a flash of insight during a sudden fit of malarial fever. In June of 1858, Wallace sent Darwin a brief essay on his views. The essay was entitled *On the Tendencies of Varieties to Depart Indefinitely from the Original Type.* With the receipt of this essay, Darwin was then induced to make a statement of his own with that of Wallace.

Wallace's essay and a portion of Darwin's manuscript, each containing remarkably similar views, were read simultaneously before the Linnaean Society in London on July 1, 1858. The joint reading of the papers stirred little interest. Darwin then labored for eight months to compress his voluminous notes into a single book, which he modestly called "only an Abstract." Wallace shares with Darwin the honor of establishing the mechanism by which evolution is brought about, but it was the monumental *The Origin of Species,* with its impressive weight of evidence and argument, that left its mark on mankind. The full title of Darwin's treatise was *On the Origin of Species by Means of Natural Selection, or the Preservation of Favoured Races in the Struggle for Life.* The first edition, some 1,250 copies, was sold out on the very day it appeared, November 24, 1859. The book was immediately both acidly attacked and effusively praised. Today, *The Origin of Species* remains the one book to be read by all serious students of nature.

The idea of evolution—that organisms change—did not originate with Darwin. There were many before him, notably Lamarck, Georges Buffon, and his grandfather, Erasmus Darwin, who recognized or intimated that animals and plants had not remained unchanged through time, but were continuously changing. Indeed, early Greek philosophers, who wondered about everything, speculated on the gradual progression of life from simple to complex. It was reserved for Darwin to remove the doctrine of evolution from the domain of speculation. Darwin's outstanding achievement was his discovery of the principle of natural selection. In showing *how* evolution occurs, Darwin endeavored to convince skeptics that evolution *does* occur.

As a whole, the principle of natural selection stems from three important observations and two deductions that logically follow from them. The first observation is that all living things tend to increase their numbers at a prolific rate. A single oyster may produce as many as 100 million eggs at one spawning; one tropical orchid may form well over 1 million seeds; and a single salmon can deposit 28 million eggs in one season. It is equally apparent (the second observation) that no one group of organisms swarms uncontrollably over the surface of the earth. In fact, the actual size of a given population of any particular organism remains relatively constant over long periods of time. If we accept these readily confirmable observations, the conclusion necessarily follows that not all individuals that are produced in any generation can survive. There is inescapably in nature an intense "struggle for existence."

Darwin's third observation was that individuals in a population are not alike but differ from one another in various features. That all living things vary is indisputable. Those individuals endowed with the most favorable variations, concluded Darwin, would have the best chance of surviving and passing their favorable characteristics on to their progeny. This differential survival, or "survival of the fittest," was termed *natural selection*. It was the British philosopher Herbert Spencer who proposed the expression "survival of the fittest," which Darwin accepted as equivalent to natural selection. In fact, Spencer suggested that Darwin had discovered not merely the laws of biological evolution but also those governing human societies.

Darwin presents the essence of his concept of natural selection in the introduction to the *Origin of Species,* as follows:

> As many more individuals of each species are born than can possibly survive; and as, consequently, there is a frequently recurring struggle for existence, it follows that any being, if it vary however slightly in any manner profitable to itself, under the complex and sometimes varying conditions of life, will have a better chance of surviving, and thus be *naturally selected.* From the strong principle of inheritance, any selected variety will tend to propagate its new and modified form.

Differential Reproduction

The survival of favorable variants is one facet of the Darwinian concept of natural selection. Equally important is the corollary that unfavorable variants do not survive and multiply. Consequently, natural selection necessarily embraces two aspects, as inseparable as the two faces of the same coin: the negative (elimination of the unfit) and the positive (perpetuation of the fit). In its negative role, natural selection serves as a conservative or stabilizing force, pruning out the aberrant forms from a population.

The superior, or fit, individuals are popularly extolled as those who emerge victoriously in brutal combat. Fitness has often been naively confused with physical, or even athletic, prowess. This glorification is traceable to such seductive catch phrases as the "struggle for existence" and the "survival of the fittest." But what does fitness actually signify?

The true gauge of fitness is not merely survival, but the organism's capacity to leave offspring. An individual must survive to reproduce, but not all individuals that survive do, or are able to, leave descendants. We have seen that the multilegged bullfrogs were not successful in propagating themselves. They failed to make a contribution to the next or succeeding generations. Therefore, they were unfit. Hence, individuals are biologically unfit if they leave no progeny. They are also unfit if they do produce progeny, none of whom survives to maturity. The less spectacular normal-legged frogs did reproduce, and to the extent that they are represented by descendants in succeeding generations, they are the fittest. Fitness, therefore, is measured as reproductive effectiveness. Natural selection can thus be thought of as *differential reproduction,* rather than differential survival.

Evolution Defined

Any given generation is descended from only a small fraction of the previous generation. It should be evident that the genes transmitted by those individuals who most successfully reproduce will predominate in the next generation. Because of unequal reproductive capacities of individuals with different hereditary constitutions, the genetic characteristics of a population become altered each successive generation. This is a dynamic process that has occurred in the past, occurs today, and will continue to occur as long as inheritable variation and differing reproductive abilities exist. Under these circumstances, the composition of a population can never remain constant. This, then, is evolution—*changes in the genetic composition of a population with the passage of each generation.*

The outcome of the evolutionary process is adaptation of an organism to its environment. Many of the structural features of organisms are marvels of construction. It is, however, not at all remarkable that organisms possess particular characteristics that appear to be precisely and peculiarly suited to their needs. This is understandable because the individuals that leave the most descendants are most often those that are best equipped to cope with special environmental conditions. In other words, the more reproductively fit individuals tend to be those that are better adapted to the environment.

An important consideration is that evolution is a property of populations. A population is not a mere assemblage of individuals; it is a breeding community of individuals. Moreover, it is the *population* that evolves in time, *not* the *individual.* An individual cannot evolve, for he or she survives for only one generation. An individual, however, can ensure continuity to a population by leaving offspring. Life is perpetuated only by the propagation of new individuals who differ in some degree from their parents. Natural selection is a term serving to inform us that some individuals leave more offspring than others. Natural selections is not purposeful or guided by a specific aim; it does not seek to attain a specific end. Natural selection is solely the consequence of the differences between individuals with respect to their capacity to produce progeny.

Throughout the ages, appropriate adaptive structures have arisen as the result of gradual changes in the hereditary endowment of a population. Admittedly, past events are not amenable to direct observation or experimental verification. There are no living eyewitnesses of very distant events. So, the process of evolution in the past must be inferred. Nevertheless, we may be confident that the same evolutionary forces we witness in operation today have guided evolution in the past. The basis for this confidence is shown in the chapters ahead.

Selected Readings

Darwin, C. 1967. *On the origin of species* (1859). New York: Atheneum Publishers.

————. 1962. *The voyage of the Beagle* (1840). Garden City, N.Y.: Doubleday & Co.

Dobzhansky, T. 1962. *Mankind evolving: the evolution of the human species*. New Haven, Conn.: Yale University Press.

Dobzhansky, T.; Ayala, F. J.; Stebbins, G. L.; and Valentine, J. M. 1977. *Evolution*. San Francisco: W. H. Freeman and Co.

Eiseley, L. 1957. *The immense journey*. New York: Random House.

————. 1958. *Darwin's century*. Garden City, N.Y.: Doubleday & Co.

Greene, J. C. 1961. *The death of Adam*. New York: The New American Library.

Irvine, W. 1972. *Apes, angels, and victorians*. New York: McGraw-Hill Book Co.

Lerner, I. M., and Libby, W. J. 1976. *Heredity, evolution, and society*. San Francisco: W. H. Freeman and Co.

Mayr, E. 1978. Evolution. *Scientific American*, September, pp. 46-55.

Moore, R., and editors of Time-Life Books. 1964. *Evolution*. New York: Time-Life Books.

Moorehead, A. 1969. *Darwin and the Beagle*. New York: Harper & Row.

Simpson, G. G. 1951. *The meaning of evolution*. New York: The New American Library.

————. 1966. *This view of life*. New York: Harcourt, Brace and World.

Stansfield, W. D. 1977. *The science of evolution*. New York: Macmillan Co.

Stebbins, G. L. 1971. *Processes of organic evolution*. Englewood Cliffs, N.J.: Prentice-Hall.

2
Inheritable Variation

Darwin recognized that the process of evolution is inseparably linked to the mechanism of inheritance. But he could not explain satisfactorily how a given trait is transmitted from parent to offspring, nor could he account adequately for the sudden appearance of new traits. Unfortunately, Darwin was unaware of the great discovery in heredity made by a contemporary, the humble Austrian monk Gregor Johann Mendel.

Mendel's manuscript appeared in print in 1866. But his published paper was overlooked by most scientists of the day, including Darwin. The distinguished German botanist Karl von Nägeli corresponded with Mendel, but he surprisingly failed to grasp that Mendel had formulated the fundamental laws of inheritance. Mendel's valuable contribution lay ignored or unappreciated until 1900, sixteen years after his death. The rediscovery of Mendel's publication at the beginning of the twentieth century ushered in the science of genetics and ultimately led to our current understanding of inheritable variation.

Mendel's Law of Segregation

Mendel's profound inference was that traits are passed on from parent to offspring through the gametes, in specific discrete factors, or units. The individual units do not blend and do not contaminate one another. The hereditary units of the parents can, therefore, be reassorted in varied combinations in different individuals at each generation. Today, Mendel's hereditary units are called *genes*. The almost infinite diversity that exists among individuals is attributable largely to the shuffling of tens of thousands of discrete genes that occurs in sexual reproduction.

The units of heredity, or genes, typically occur in pairs in the fertilized egg. The offspring inherits one gene of each pair from the female parent and one of each pair from the male parent. In the most elementary case of inheritance, the trait is governed by a single pair of genes. Albinism in humans is an example of a trait under the simple control of one pair of genes.

The word *albino* is derived from the Latin *albus*, meaning white, and refers to the paucity or absence of pigment (melanin) in the skin, hair, and eyes. The skin is often very light ("milk-white"), and the hair whitish-yellow. The eyes appear pinkish because the red blood vessels give the otherwise colorless iris a rosy cast. Albinos have poor vision and are acutely sensitive to sunlight. They are especially prone to skin cancer. In some cultures, as among the natives of the Archipelago Coral Islands, albinos are believed to have a mesmeric aura (fig. 2.1).

Albinism results from a defective gene. Specifically, the condition arises only when the individual is endowed with two defective genes, one from each parent. Remember that the affected person must have received the defective gene for albinism from *both* parents.

Figure 2.1 Albino person showing the complete absence of pigment in the eyes, hair, and skin. Natives of the Archipelago Coral Islands (100 miles east of the New Guinea mainland) are said to take care of their "pure white" relatives who can hardly see in the glare of the tropical sun. (Wide World Photos.)

A cardinal fact of inheritance is that each gamete (egg or sperm) contains *one, and only one,* member of a pair of genes. Thus, when the egg and sperm unite, the two genes for each character are brought together in the new individual. During the production of the gametes, the members of a pair of genes separate, or *segregate*, from each other. A given gamete can carry *A* or *a*, but not both. This fundamental concept that only one member of any pair of genes in a parent is transmitted to each offspring is Mendel's first law—the *law of segregation.*

As seen in figure 2.2, there are nine possible marriages with respect to the pair of genes conditioning the presence or absence of melanin. If both parents are perfectly normal—that is, each possesses a pair of normal genes—then all offspring will be normally pigmented (cross 1). When both parents are albino and accordingly each has a pair of defective genes, all offspring will be afflicted with albinism (cross 2). The persons involved in either of these two crosses are said to be *homozygous*, since they each possess a pair of similar, or like, genes. The normal individuals in cross 1 are homozygous for the normal gene, and the albino individuals in cross 2 are homozygous for the defective gene.

Looking at cross 3, we notice that one of the parents (the mother) carries an unlike pair of genes: one member of the pair is normal; its partner gene is defective. This parent, possessing unlike genes, is *heterozygous*. This parent has normal coloration, which leads us to deduce that the expression of the defective gene is completely masked or suppressed by the normal gene. In genetical parlance, we say that the normal gene is *dominant* over its partner, the *recessive* gene. The dominant gene is customarily symbolized by a capital letter (in our case, *A*); the recessive gene, by a corresponding small letter (*a*). The heterozygous parent is depicted as *Aa*, and is referred to as *heterozygote* or a *carrier*.

When the heterozygote (*Aa*) marries a dominant homozygote (*AA*), all the offspring will be outwardly normal, but half the offspring will be carriers (*Aa*) like one of the parents (cross 3). The ratio of *AA* to *Aa* offspring may be expressed as 1/2 *AA* : 1/2 *Aa* or simply 1 *AA* : 1*Aa*. Cross 4 differs from cross 3 only in that the male parent is the carrier (*Aa*) and the female parent is the dominant homozygote (*AA*). The outcome is the same: all children are normal in appearance; but half of them are carriers. Thus, with respect to albinism, the males and females are equally likely to have or to transmit the trait. (There are traits, such as hemophilia and red-green color-blindness, in which transmission is influenced by the sex of the parents.)

All offspring will be carriers from the marriage (crosses 5 and 6) of a dominant homozygote (*AA*) and an albino parent (*aa*). Examining crosses 5 and 6 (fig. 2.2), we again notice that males and females are equally affected. The marriage of a heterozygous person (*Aa*) and an albino (*aa*) gives rise to two types of progeny, in equal numbers (crosses 7 and 8). Half the offspring are normal but carriers (*Aa*); the remaining half are albino (*aa*).

Figure 2.2 **The segregation** of a single pair of genes. The nine possible marriages with respect to the recessive trait for albinism show the operation of Mendel's law of segregation.

The outcome of the last of the crosses (no. 9) in figure 2.2 is comprehensible in terms of Mendel's law of segregation. Both parents are normal in appearance, but each carries the defective gene (a) masked by the normal gene (A). The male parent (Aa) produces two kinds of sperm cells: half the sperm carry the A gene; half carry the a allele. The same two kinds occur in equal proportions among the egg cells. Each kind of sperm has an equal chance of meeting each kind of egg. The random meeting of gametes leads to both normal and albino offspring. One-fourth of the progeny are normal and completely devoid of the recessive gene (AA); one-half are normal but carriers like the parents (Aa); and the remaining one-fourth exhibit the albino anomaly (aa). The probabilities of occurrence are the same as those illustrated by the tossing of a coin. If two coins are repeatedly tossed at the same time, the result to be expected on the basis of pure chance is that the combination "head and head" will fall in one-fourth of the cases, the combination "head and tail" in one-half of them, and the combination "tail and tail" in the remaining one-fourth. The ratio of genetic constitutions among the offspring may be expressed as 1/4 AA : 1/2 Aa : 1/4 aa, or 1 AA : 2 Aa : 1 aa.

Geneticists constantly have to make a distinction between the external, or observable, appearance of the organism and its internal hereditary constitution. They call the former the *phenotype* (visible type) and the latter the *genotype* (hereditary type). In our last cross (no. 9), there are three different genotypes: AA, Aa, and aa. The dominant homozygote (AA) cannot be distinguished by inspection from the heterozygote (Aa). Thus, AA and Aa genotypes have the same phenotype. Accordingly, on the basis of phenotype alone, the progeny ratio is 3/4 normal : 1/4 albino, or 3 normal : 1 albino.

If we disregard the sexes of the parents, the nine crosses in figure 2.2 become reducible to six possible types of matings. The six kinds of matings involving a single pair of genes are set forth in table 2.1.

Application of Mendelian Principles

Our interpretation of the mode of inheritance of the multilegged trait in the bullfrog was based on Mendel's principles. In the preceding chapter, we postulated a cross between two normal, but heterozygous, frogs. The expectation for multilegged offspring from two heterozygous parents is 25 percent, in keeping with Mendel's 3:1 ratio. This cross, using appropriate genetic symbols, is shown graphically by "checkerboard" analysis in figure 2.3. The convenient checkerboard method of analysis was devised by the British geneticist R. C. Punnett. The eggs and sperm are listed separately on two different sides of the checkerboard, and each square represents an offspring that arises from the union of a given egg cell and a given sperm cell. The "reading" of the squares discloses the classical Mendelian phenotypic ratio of 3:1.

Table 2.1

Simple Recessive Mendelian Inheritance, Involving a Single Pair of Genes (Normal Pigmentation vs. Albinism)

MATING TYPES		GAMETES		OFFSPRING	
Genotypes	Phenotypes	First Parent 50% 50%	Second Parent 50% 50%	Genotypes	Phenotypes
$AA \times AA$	Normal × Normal	A A	A A	100% AA	100% Normal
$AA \times Aa$	Normal × Normal	A A	A a	50% AA 50% Aa	100% Normal
$Aa \times Aa$	Normal × Normal	A a	A a	25% AA 50% Aa 25% aa	75% Normal 25% Albino
$AA \times aa$	Normal × Albino	A A	a a	100% Aa	100% Normal
$Aa \times aa$	Normal × Albino	A a	a a	50% Aa 50% aa	50% Normal 50% Albino
$aa \times aa$	Albino × Albino	a a	a a	100% aa	100% Albino

It should be understood that the 3:1 phenotypic ratio resulting from the mating of two heterozygous persons, or the 1:1 ratio from the mating of a heterozygote and a recessive, are expectations based on probability, and not invariable outcomes. The production of large numbers of offspring increases the probability of obtaining, for example, the 1:1 progeny ratio, just as many tosses of a coin improve the chances of approximating the 1 head : 1 tail ratio. If a coin is tossed only two times, a head on the first toss is not invariably followed by a tail on the second toss. In like manner, if only two offspring are produced from a marriage of a heterozygote and a recessive parent, it should not be thought that one normal offspring is always accompanied by one recessive offspring. With small numbers of progeny, as is characteristic of humans, any ratio might arise in a given family.

Stated another way, the 3:1 and 1:1 ratios reveal the risk or odds of a given child having the particular trait involved. For example, if the first child of two heterozygous parents is an albino, the odds that the second child will be an albino remain 1 out of 4. These odds hold for each subsequent child, irrespective of the number of previously affected children. Each conception is an entirely independent event. If this still appears puzzling, consider once again the tossing of a coin. The first time a coin is tossed, the chance of obtaining either a head or a tail is 1 in 2. Whether the toss is repeated immediately or nine months later, the chance of the coin falling head or tail is still 1 in 2, or 50 percent.

Figure 2.3   Cross of two heterozygous carrier frogs. One-fourth of the offspring are normal and completely devoid of the recessive gene (*AA*); one-half are normal but carriers, like the parents (*Aa*); and the remaining one-fourth exhibit the multilegged anomaly (*aa*).

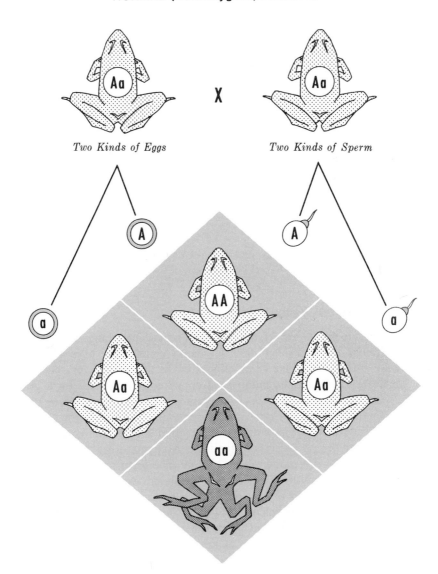

NORMAL (Heterozygous) PARENTS

3/4 NORMAL : 1/4 MULTILEGGED

Figure 2.4   **Normal human complement** of 46 chromosomes, prepared from a culture of blood cells of a male. Each chromosome consists of two strands. The chromosomes occur in pairs; one member of each chromosome pair is derived from the mother and the other is derived from the father. Thus, each chromosome has a homologous ("like") mate in the complement.

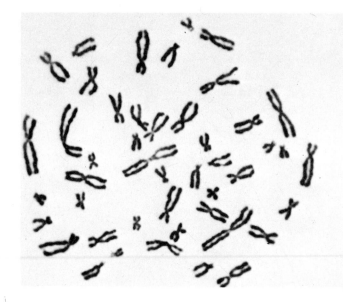

Chromosomal Basis of Heredity

Each organism has a definite number of chromosomes, ranging in various species from 2 to more than 1,000. Humans, for example, have 46 microscopically visible, threadlike chromosomes in the nucleus of every body cell (fig. 2.4). Not only are the numbers of chromosomes constant for a given species, but the shapes and sizes of the chromosomes in a given species differ enough for us to recognize particular chromosomes and to distinguish one kind from the other. An important consideration is that each kind of chromosome in a zygote is present in duplicate. Stated another way, the chromosomes occur in pairs; humans have 23 pairs. The significance of the double, or paired, set of chromosomes, cannot be overstated: *each parent contributes one member of each pair of chromosomes to the offspring.* The behavior of chromosomes parallels the behavior of genes. Mendel's findings revealed that the genes occur in pairs, and that the members of each pair of genes separate from each other in the production of the gametes. Evidently then, we may deduce that the genes are located in the chromosomes. Extensive studies have provided strong confirmation that the chromosomes do indeed carry the genes.

Two chromosomes that form a pair are *homologous.* A special kind of division guarantees that a gamete contains one member of homologous pairs of chromosomes. This special process is called *meiosis.* Meiosis is a lessening or reducing process, referring specifically to a reduction in the number of chromosomes. In humans, for example, during formation of the gametes, the number of chromosomes is reduced from 46 to 23.

Figure 2.5 shows the separation of the pair of homologous chromosomes during meiosis of the heterozygote carrying the *A* and *a* genes. The gene *A* occupies a particular site, or *locus*, in one of the chromosomes. The alternative gene, *a*, occurs at the identical locus in the other homologous chromosome. The alternative forms of a given gene, occupying a given locus, are termed *alleles*. Thus, *A* is an allele of *a* (or, *A* is allelic to *a*). Now, we have earlier seen that the heterozygote (*Aa*) produces two types of gametes, *A* and *a*, in equal numbers. It should be evident from figure 2.5 that the gametic ratio of 1 *A* : 1 *a* is the consequence of the separation of the two homologous chromosomes during meiosis, one chromosome containing the *A* gene and its homologue carrying the *a* allele. The behavior of chromosomes also explains another feature of Mendel's results—the independent assortment of traits.

Mendel's Law of Independent Assortment

The fruit fly (*Drosophila melanogaster*), approximately one-fourth the size of a house fly, has long been a favorite subject for genetical research. Numerous variations have been found, affecting all parts of the body. For example, the wings may be reduced in length to small vestiges, the result of action of a recessive, *vestigial* gene. Two heterozygous normal-winged flies, each harboring the recessive gene for vestigial wings, can produce two types of offspring: normal-winged and vestigial-winged, in a ratio of three-to-one. Now suppose that these same parents have a second concealed recessive gene, namely, *ebony*. This recessive gene, when homozygous, modifies the normal gray color of the body to black, or ebony.

Two normal-winged, gray-bodied parents, each heterozygous (*NnGg*), can give rise to four types of offspring phenotypically, as seen in figure 2.6. Two of the phenotypes are like the original parents, and two are new combinations (normal-winged, ebony and vestigial-winged, gray). The four phenotypes appear in a ratio of 9:3:3:1. For such a ratio to be obtained, each parent (*NnGg*) must have produced four kinds of gametes in equal proportions: *NG, Ng, nG,* and *ng*. Stated another way, the segregation of the members of one pair of genes must have occurred independently of the segregation of the members of the other gene pair during gamete formation. Thus, 50 percent of the gametes received *N*, and of these same gametes, half obtained *G* and the other half *g*. Accordingly, 25 percent of the gametes were *NG* and 25 percent were *Ng*. Likewise, 50 percent of the gametes carried *n*, of which half contained as well *G* and half *g* (or 25 percent *nG* and 25 percent *ng*). The four kinds of egg cells and the four kinds of sperm cells can unite in 16 possible ways, as shown graphically in the "checkerboard" in figure 2.6.

The above cross illustrates Mendel's *law of independent assortment*, which states that one trait (one gene pair) segregates independently of another trait (another gene pair). The separation of the *Nn* pair of genes and the simultaneous independent separation of the *Gg* pair of genes occurs because the two pairs of genes are located in two different pairs of chromosomes. Once again, the parallelism in the behavior of genes and chromosomes is striking.

Figure 2.5 Chromosomal basis of inheritance. When one chromosome of a pair carries a given gene (*A*, in this case), and its homologue carries the alternative form of the gene *(a)*, then meiosis results in the production of two distinct kinds of gametes in equal proportions, *A* and *a*. The scheme of events presented here is highly simplified.

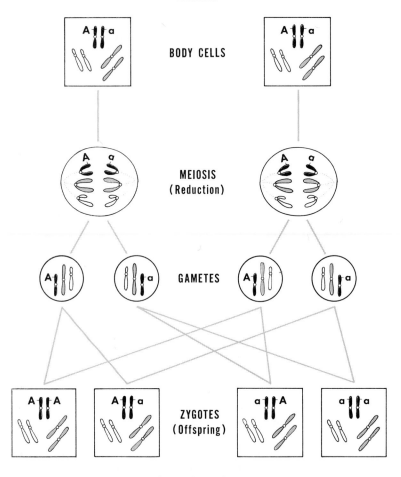

PARENTS

BODY CELLS

MEIOSIS (Reduction)

GAMETES

ZYGOTES (Offspring)

1AA : 2Aa : 1aa

Three or More Gene Pairs

When individuals differing in three characteristics are crossed, the situation becomes more complex, but the principle of independent assortment still holds. An individual heterozygous for three independently assorting pairs of alleles (for example, *Aa, Bb, and Cc*) produces eight different types of gametes in equal numbers, as illustrated in figure 2.7. Random union among these eight kinds of gametes yields 64 equally possible combinations. Stated another way,

Figure 2.6 Mendel's law of independent assortment, as revealed in the inheritance of two pairs of characters in the fruit fly. A pure normal-winged fly with gray body mated with a vestigial-winged, ebony-bodied fly produces all normal-winged, gray-bodied flies in the first generation (F_1). When these F_1 flies are inbred, a second generation (F_2) is produced that displays a phenotypic ratio of 9:3:3:1.

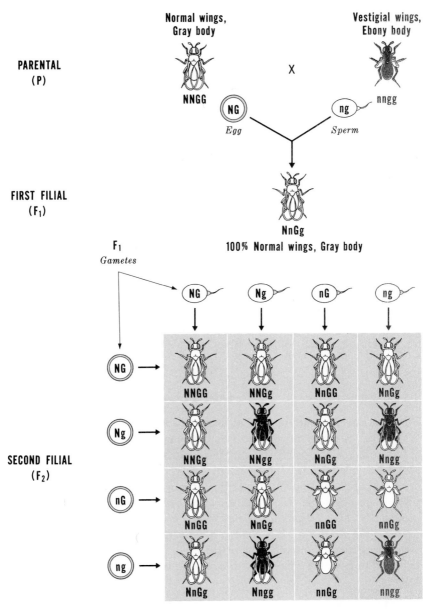

Figure 2.7 Gamete formation in an individual heterozygous for three pairs of genes ("trihybrid"). The members of the three different pairs of genes assort independently of each other when gametes are formed. The separation of the *Aa* pair, *Bb* pair, and the *Cc* pair, respectively, can be treated as independent events. The outcome is eight genetically different gametes.

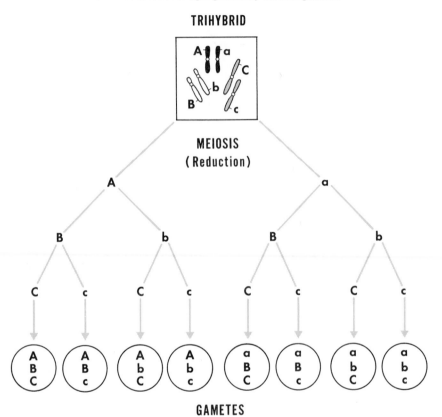

Punnett's checkerboard method of analysis would have 64 (8 × 8) squares. An analysis of the offspring reveals that, among the 64 possible combinations, there are only eight visibly different phenotypes. You may wish to verify that the phenotypic ratio is 27:9:9:9:3:3:3:1.

The use of the Punnett square for three or more independently assorting genes is cumbersome. Here we will introduce one of the cardinal rules of probability: *The chance that two or more independent events will occur together is the product of their chances of occurring separately.* As an example, what proportion of the offspring of the cross *AaBbCc* × *AaBbCc* would be expected to have the genotype *AaBBcc*? The individual computations are as follows:

1. The chance that an individual will be *Aa* = 1/2.
2. The chance that an individual will be *BB* = 1/4.
3. The chance that an individual will be *cc* = 1/4.

Assuming independence among these three pairs of alternatives, the chance of the three combining together (*AaBBcc*) is the product of their separate probabilities, as follows:

$$1/2 \times 1/4 \times 1/4 = 1/32.$$

Significance of Independent Assortment

The Mendelian principles permit a genuine appreciation of the source of inheritable variation in natural populations of organisms. An impressively large array of different kinds of individuals can arise from the segregation and reassortment of independent pairs of genes. Figure 2.8 considers a few of the independently assorting traits in the fruit fly. It is seen that the number of visibly different classes (phenotypes) of offspring is doubled by each additional heterozygous pair of independently assorting genes. Each additional heterozygous gene pair multiplies the number of visibly different combinations by

Figure 2.8 **Inheritance of traits in the fruit fly,** when the parents are each heterozygous for one (*Aa*), two (*AaBb*), and three (*AaBbCc*) independently assorting traits. The gene for long wings (*A*) is dominant over the gene for vestigial wings (*a*); gray body (*B*) is dominant over ebony (*b*); and normal eyes (*C*) is dominant over the eyeless condition (*c*). The number of visibly different classes (phenotypes) of offspring increases progressively with each additional pair of traits. The generalized formula for determining the number of different phenotypes among the offspring (when dominance is complete) is 2^n, where *n* stands for the number of different heterozygous pairs of genes.

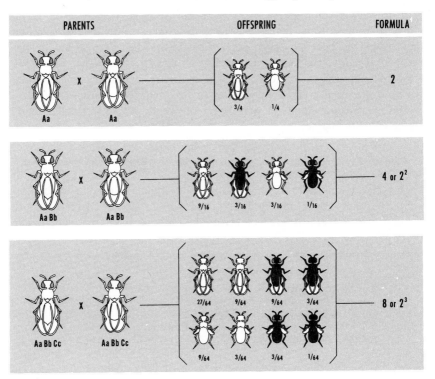

PARENTS	OFFSPRING	FORMULA
Aa X *Aa*	3/4 · 1/4	2
Aa Bb X *Aa Bb*	9/16 · 3/16 · 3/16 · 1/16	4 or 2^2
Aa Bb Cc X *Aa Bb Cc*	27/64 · 9/64 · 9/64 · 3/64 · 9/64 · 3/64 · 3/64 · 1/64	8 or 2^3

Table 2.2
Characteristics of Crosses Involving Various Pairs of Independently Assorting Genes

Number of heterozygous pairs involved in the cross	Number of different types of gametes produced by the heterozygote	Number of visibly different progeny (that is, phenotypes)*
1	2	2
2	4	4
3	8	8
4	16	16
10	1,024	1,024
n	2^n	2^n

*Assumes complete dominance in each pair.

a factor of two. Thus, if the parents are each heterozygous for ten pairs of genes (when dominance is complete), the number of different phenotypes among the progeny becomes 2^{10}, or 1,024.

Other mathematical regularities are evident as we increase the number of heterozygous pairs (table 2.2). The number of different genotypes among the progeny is a multiple of the base 3, that is, 3^n. Moreover, as the number of heterozygous traits increases, the chance of recovering one of the homozygous classes becomes progressively less. Thus, when a single heterozygous pair (Aa) is involved, one in four will be either AA or aa. When four heterozygous traits are involved, only one in 256 will be either homozygous dominant or recessive for all traits. Evidently, no single genetic constitution is ever likely to be exactly duplicated in a person (save in identical twins). Each individual is genetically unique.

The process of independent combination of genes may be compared, on a very modest scale, to the shuffling and dealing of playing cards. One pack of 52 cards can yield a large variety of hands. And, just as in poker a full house is far superior to three of a kind, so in organisms certain combinations of genes confer a greater reproductive advantage to its bearer than other combinations. Natural selection favors the more reproductive genotypes.

Selected Readings

Beadle, G. W., and Beadle, M. 1968. *The language of life*. Garden City, N.Y.: Doubleday & Co.

Gardner, E. J. 1968. *Principles of genetics*. New York: John C. Wiley & Sons.

Number of genotypically different combinations (that is, genotypes)	Chance of recovering an individual that is either homozygous dominant or recessive for all traits
3	1 in 4
9	1 in 16
27	1 in 64
81	1 in 256
59,049	1 in 1,048,576
3^n	1 in 4^n

Goodenough, U., and Levine, R. P. 1978. *Genetics.* New York: Holt, Rinehart & Winston.

Iltis, H. 1932. *Life of Mendel.* New York: W. W. Norton & Co.

Jenkins, J. B. 1979. *Genetics.* Boston: Houghton Mifflin Co.

McKusick, V. A. 1969. *Human genetics.* Englewood Cliffs, N.J.: Prentice-Hall.

Mendel, G. 1948. *Experiments in plant hybridization.* Cambridge, Mass.: Harvard University Press. (Translation of Gregor Mendel's original paper first published in 1866).

Moore, J. A. 1972. *Heredity and development.* New York: Oxford University Press.

Pai, A. C. 1974. *Foundations of genetics.* New York: McGraw-Hill Book Co.

Peters, J. A. 1959. *Classic papers in genetics.* Englewood Cliffs, N.J.: Prentice-Hall.

Scheinfeld, A. 1965. *Your heredity and environment.* Philadelphia: J. B. Lippincott Co.

Stern, C. 1973. *Principles of human genetics.* San Francisco: W. H. Freeman and Co.

Stern, C., and Sherwood, E. R. 1966. *The origin of genetics.* San Francisco: W. H. Freeman and Co.

Stine, G. J. 1977. *Biosocial genetics.* New York: Macmillan Co.

Strickberger, M. W. 1976. *Genetics.* New York: Macmillan Co.

Sturtevant, A. H. 1965. *A history of genetics.* New York: Harper & Row.

Watson, J. D. 1976. *Molecular biology of the gene.* New York: W. A. Benjamin.

3
Mutation

In the opening pages of *The Origin of Species,* Charles Darwin declared, "Any variation which is not inherited is unimportant for us." This and other incisive statements show that Darwin understood the importance of inherited variability. He was, however, eternally assailed by doubts as to how new varieties initially arise. Today we know that the underlying cause of variation is *mutation.*

Each gene of an organism may assume a variety of forms. A normal gene may change to another form and produce an effect on a trait different from that of the original gene. An inheritable change in the structure of the gene is known as mutation. Variant traits, such as albinism in humans and vestigial wings in the fruit fly, are traceable to the action of altered, or mutant, genes. Mutations are the ultimate source of genetic variation. *All differences in the genes of organisms have their origin in mutation.*

Causes of Mutation

New mutations arise from time to time, and the same mutations may occur repeatedly. It is often difficult to distinguish between new mutations and old ones that occurred previously and were carried concealed in ancestors. A recessive mutant gene may remain masked by its normal dominant allele for many generations, and reveal itself for the first time only when two heterozygous carriers of the same mutant gene happen to mate.

Each gene runs the risk of changing to an alternative form. The causes of naturally occurring, or *spontaneous,* mutations are largely unknown. The environment contains a background of inescapable radiation from radioactive elements, cosmic rays, and gamma rays. It is generally conceded that the amount of background radiation is too low to account for all spontaneous mutations. In other words, only a small fraction of spontaneous mutations can be attributed to background radiation.

In 1927, the late Nobel laureate Hermann J. Muller of Indiana University announced that genes are highly susceptible to the action of X rays. By irradiating fruit flies with X rays, he demonstrated that the process of mutation is enormously speeded up. The production of mutations is directly proportional to the total dosage of X rays. Stated another way, the greater the dosage, the greater the mutation rate. In this linear, or straight-line, relationship, the mutation rate increases proportionately about 3 percent for each 1,000 units (or *roentgens*) of X rays.

It has long been held that the mutagenic effect is the same whether the dose is given in a short time or spread over a long period. In other words, low intensities of X rays over long periods of time produce as many mutations as the same dose administered in high intensities in a short period of time. Recent experiments on mice have cast some doubt on this view, for it has been shown (at least in mice) that the mutagenic effect of a single exposure to the germ cells is greater than the effect of the same exposure administered as several smaller doses separated by intervals of time. Moreover, there does not appear to be a critical, or *threshold,* dose of roentgens below which there is no effect. In essence, no dose is so low (or "safe") as to carry no risk of inducing a mutation. Modern workers stress that any amount of radiation, no matter how little, can cause a mutation.

At the time of Muller's discovery, no one conceived that within a generation the entire population of humans would be exposed to a significant increase of high-energy radiation as a consequence of the creation of the atomic bomb. The additional amount of high-energy radiation already produced by fallout from atomic explosions has most likely increased the mutation rate. Most of the radiation-induced mutations are recessive and most of them are deleterious.

In discussing the hazards of X rays and other ionizing radiations, we must be careful to distinguish between *somatic damage* and *genetic damage.* Injury to the body cells of the exposed individuals themselves constitutes somatic damage. On the other hand, impairment of the genetic apparatus of the sex cells represents genetic damage. Typically, the genetic alterations do not manifest themselves in the individuals themselves but present a risk for their descendants in the immediate or succeeding generations.

There are documented records of the *somatic* consequences of exposure to the 1945 atomic blasts in Hiroshima and Nagasaki. Among 161 children born of women who were exposed to the atomic bomb while pregnant, 29 were microcephalic (head size considerably below normal) and 11 of these 29 were mentally retarded. As might have been expected, the detrimental effects were most pronounced among the infants of women who were in an early stage of pregnancy (less than 15 weeks' gestation) at the time of exposure. Moreover, most of the women who gave birth to deformed infants were less than 1.3 miles from the center (hypocenter) of the explosion. The adverse effects on the fetus diminished in frequency and severity as the distance from the hypocenter increased.

Analyses have also revealed that survivors of the atomic blasts have developed leukemia in rough proportion to the dosage of ionizing radiation. From 0.02 to 1.00 percent of the survivors within the range of radiation developed leukemia in the decade between 1950 and 1960, with the greatest percentage of cases present in individuals nearest the hypocenter. The data substantiate the association found in other studies between whole-body exposure to radiation at high dose levels and the incidence of leukemia.

Estimates of the magnitude of possible *genetic* damages have been of uncertain significance. The statistical methods used are too insensitive to detect the occurrence of radiation-induced mutations. There was no greater amount of gene-determined defects to infants born to women who were pregnant at the time of the atomic blasts at Hiroshima and Nagasaki compared with populations that were not exposed. However, it must be stressed that most of the possible recessive mutations induced would not be expressed for several generations.

In an effort to measure immediate damaging genetic effects, a possible shift in the sex ratios among children of exposed parents was sought. Theoretically, if fatal, or lethal, mutations are induced on the X chromosome, they will be immediately expressed in the sons of the exposed mother (fig. 3.1). Deaths of male fetuses prior to birth can be indirectly gauged by a decrease in the number of males born to irradiated mothers. An early, limited study on

Figure 3.1   Death of male fetuses resulting from radiation-induced lethal mutations in one of the X chromosomes of the mother. Theoretically, the sex ratio (the proportion of males to females) among children conceived after exposure of the expectant mother to radiation should be diminished.

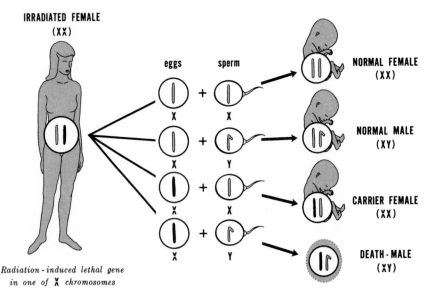

the offspring of Japanese mothers who received a heavy dose of radiation revealed a slight reduction of male births, but a later study, based on a larger sample, showed no alteration of the sex ratio.

Although the actual data available have failed to reveal unequivocal genetic effects of radiation, it would be a patent fallacy to conclude that atomic radiation has had no mutagenic effect. American and British geneticists estimate that each person currently receives a total dose of 7.8 roentgens to the reproductive cells during the first 30 years of life. Of this amount, 3.1 roentgens are derived from natural background radiation, 4.6 roentgens from various medical uses of ionizing radiation, and an additional 0.1 roentgen from the testing of nuclear weapons. The additional amount of radiation (0.1 roentgen) received from nuclear fallout may seem trivial. However, the exposure of our population to 0.1 roentgen is calculated to induce sufficient mutations to result in 3,750 defective offspring among 100 million births.

Chemical Nature of the Gene

During the early decades of this century, the picture of genes that had emerged was that of discrete entities, strung together along the chromosome, like individual beads on a string. This portrayal carried with it the implication that genes were not pure abstractions, but actual particles of living matter. Yet, until the 1950s the chemical nature of the gene remained enigmatic. Also, no one knew precisely how genes made additional copies of themselves or how they controlled the development of a trait.

One of the finest triumphs of modern science has been the elucidation of the chemical nature of the gene. The transmission of traits from parents to offspring depends on the transfer of a specific giant molecule that carries a coded blueprint in its molecular structure. This complex molecule, the basic chemical component of the chromosome, is *deoxyribonucleic acid* (DNA). The information carried in the DNA molecule can be divided into a number of separable units, which we now recognize as the genes. Stated simply, chromosomes are primarily long strands of DNA, and genes are coded sequences of the DNA molecule. The amount of DNA present in a single unfertilized human egg has been estimated to carry information corresponding to two million genes.

The two scientists who worked together in the early 1950s to propose a configuration for the DNA molecule were Francis H. C. Crick, a biophysicist at Cambridge University, and James D. Watson, an American student of virology who was then studying chemistry at Cambridge on a postdoctoral fellowship. With the invaluable aid of X-ray pictures of DNA crystals prepared by Maurice H. F. Wilkins, a biophysicist at King's College in London, Watson and Crick built an inspired model in metal of DNA's configuration. This achievement won Watson, Crick, and Wilkins the coveted Nobel Prize for physiology and medicine in 1962.

Figure 3.2   **Watson-Crick double-stranded helix** configuration of deoxyribo-nucleic acid (DNA). The backbone of each twisted strand consists of alternating sugar (S) and phosphate (P) residues. Enlarged view on right shows that the larger, two-ring purines (adenine or guanine) lay opposite the smaller, one-ring pyrimidines (cytosine or thymine). The nitrogenous bases are held together by hydrogen bonds, three between cytosine (C) and guanine (G), and two between adenine (A) and thymine (T).

The remarkable feature of DNA is its simplicity of design. The double-stranded DNA molecule is shaped like a twisted ladder (fig. 3.2). The two parallel strands of the ladder are twisted around each other somewhat like the supporting frameworks of a spiral staircase. The twisted supports of the ladder are made up of alternating units of sugar (deoxyribose) and phosphate molecules, while the cross-links, or rungs, are composed of specific nitrogen-containing ring compounds, or nitrogenous bases. There are two classes of nitrogenous bases, the larger, two-ring purines and the smaller, one-ring pyrimidines. Each rung of the ladder consists of one purine coupled to one pyrimidine. The arrangement of the bases is not haphazard: adenine (A) in one chain is normally joined with thymine (T) in the other chain, and guanine

(G) is typically linked with cytosine (C). Along any one chain, any sequence of the bases is possible, but if the sequence along one chain is given, then the sequence along the other is automatically determined, because of the precise pairing rule (A=T and G=C). The combination of one purine and one pyrimidine to make up each cross connection is conveniently called a *base pair*.

The DNA molecule has a structure that is sufficiently versatile to account for the great variety of different genes. The four bases (A, C, G, and T) may be thought of as a four-letter alphabet, or code. A given gene owes its unique character to a specific order of the bases, just as words in our language differ according to the sequence of letters of the alphabet or as a telegraphic message becomes comprehensible by the varied combinations of dots and dashes. The number of ways in which the nitrogenous bases can be arranged in the DNA molecule is exceedingly large. Current estimates indicate that a single gene is a linear sequence of approximately 1,500 base pairs.

Self-Copying of DNA Molecule

One fundamental property that has long been ascribed to the gene is its ability to make an exact copy of itself. If the gene is really a linear sequence of base pairs, then the Watson-Crick model of DNA must be able to account for the precise reproduction of the sequences of the bases. The exact replication of the double helix can be visualized with little difficulty. The DNA molecule consists of two parts, each of which is the complement of the other. Accordingly, each single chain can serve as a template to guide the formation of a new companion chain (fig. 3.3). The two parallel chains separate, breaking the hydrogen bonds that hold together the paired bases. Each chain then attracts new base units from among the supply of free units always present in the cell. As separation takes place, each separated lengthwise portion of the chain can begin to form a portion of the new chain. Eventually, the original double helix has produced two exact replicas of itself. If the original two chains are designated *A* and *B,* then *A* will direct the formation of a new *B* and the old *B* will guide the production of a new *A*. Where one *AB* molecule existed previously, two *AB* molecules, exactly like the original, exist afterward.

The mechanism of DNA self-duplication (or replication) is in accord with our understanding of the duplication of chromosomes during the division of a cell. When the fertilized egg divides into many daughter cells, the chromosomes regularly produce copies of themselves with each cell division. As a result, each of the millions of cells of the developing embryo carries faithful copies of the chromosomes of the fertilized egg. The process by which new cells in the organism arise from preexisting cells is termed *mitosis*. The precise similarity of the new, or daughter, chromosomes to the original, or parent, chromosome is ensured by the mode of replication of the DNA molecule. When two identical molecules of DNA have been formed, one copy is passed on to each of the two daughter cells during mitosis. Thus, each daughter cell receives the same set of genetic instructions as was originally present in the parent cell.

Figure 3.3 Replication of DNA. The parental strands separate, each serving as a guide or template for the synthesis of a new daughter strand. Each new DNA molecule contains one parental and one daughter strand.

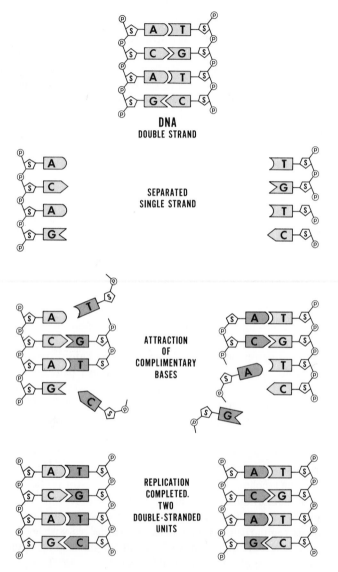

Molecular Mechanism of Mutation

DNA replication is subject to error. A mutation can be envisioned as a chance mishap to one of the paired bases in the DNA molecule. A single substitution of a C-G pair for a T-A may be sufficient to alter the character of the gene. One possible way in which such a substitution might occur is shown in figure

Figure 3.4   **Mutation at the molecular level.** When cell division occurs, the two twisted strands separate and each strand attracts unbound nucleotides (containing nitrogenous bases) to rebuild the DNA molecule. This particular illustration shows events through two successive cell divisions, commencing with strand I in the original DNA molecule. During the first division, an adenine radical (A) undergoes a chemical change to A′, which misattracts cytosine (C) instead of thymine (T). At the next division, cytosine (C) attracts its customary partner, guanine (G). The net result is that the granddaughter DNA molecule contains a C-G pair where a T-A pair was formerly located. This highly localized change in one of the pairs of nitrogenous bases may qualify as a gene mutation.

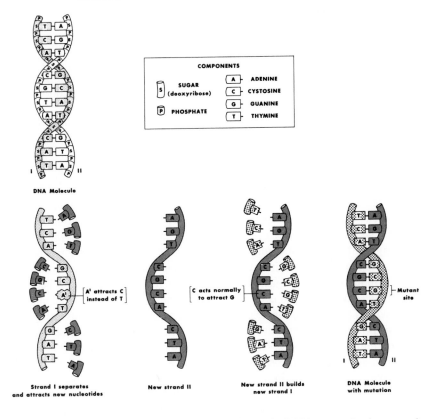

3.4. It has been established that each of the bases in DNA can exist in several rare alternative forms known as *tautomers*. The tautomeric alternatives differ in the positions at which certain atoms are attached in the molecule. The significance of the tautomeric alternatives lies in the changed pairing qualities they impart. Whereas, for example, the normal form of adenine pairs with thymine, one of its rare tautomeric forms actually attracts and couples with cytosine. Such a tautomeric shift in the adenine molecule may thus make possible a new purine-pyrimidine base pair arrangement in the DNA molecule. As seen in figure 3.4, an adenine (labeled A¹) has undergone a tautomeric shift and attracts the wrong partner (cytosine instead of thymine). At the next replication of the strand, the misattracted cytosine acts normally to join with guanine. Hence, a C-G pair is established where a T-A pair was formerly located. At this point in the DNA molecule, the gene is modified and may be expected to produce a mutant effect.

Many mutations are simply base-pair switches in the DNA molecule. Other mutational changes may involve the insertion of a new base or the deletion of one or more bases within a gene. Changes at the molecular level within the gene have been called *point* (or *gene*) *mutations*. Grosser alterations involving the chromosomes, such as the rearrangement of segments of a chromosome, have been termed *chromosomal aberrations*. Some of the chromosomal aberrations are numerical, affecting the amount of chromosomal material. A cell, for example, may have three sets of chromosomes (the *triploid* state), rather than the customary two chromosomal sets (the standard diploid state). In humans, gross chromosomal deviations from normal have been found in 5 out of every 1,000 live births and 35 of every 1,000 spontaneous abortions.

In 1942, Charlotte Auerbach and her co-workers in England reported that the World War II chemical agent mustard gas was as powerfully mutagenic as high-energy radiations. Since then, many chemical compounds once considered to be nontoxic have been shown to induce mutations in experimental animals. Some mutagenic chemicals actually resemble the normal bases in DNA to such an extent that they can substitute for the regular bases. Two prominent examples are 5-bromouracil and 2-aminopurine. Other chemical agents, such as nitrous acid and ethyl methanesulfonate, alter the gene by causing the base of replicating DNA to mispair. Nitrous acid would be a potent mutagen in nature, since it is now a common pollutant in the atmosphere.

In industrialized countries, foods in their natural states have been supplemented, modified, and even disguised by the use of innumerable chemical additives. The average American consumes about three pounds per year of chemical food additives. Several of the chemicals we ingest are suspected of having mutagenic effects. The preservative sodium nitrite has been used extensively to maintain the reddish appearance of meat. Sodium nitrite combines with the hemoglobin of raw meat to produce nitrosohemoglobin, which imparts the red color. Unfortunately, the nitroso compounds are one of the most powerful mutagenic agents known. With chemical food additives, insecticides, antibiotics, and drugs increasingly finding their way into the human diet, we may be creating a paramount problem in mutagenesis unheard of even a decade ago.

Language of Life

It is now widely accepted that genes provide the information to direct the synthesis of proteins in the cell. The gene—a linear sequence of bases in the DNA—codes for the precise sequence of amino acids that compose a protein.

The properties of any protein are determined by the number, identity, and arrangement of the amino acid residues (components). In a large sense, the molecular basis of the genotype is DNA; the molecular basis of the phenotype is protein. Ultimately all inherited differences in individuals are expressed in their bodily structure through changes in specific proteins.

Amino acids, the basic building blocks of proteins, are arranged in long chains called polypeptides. Each polypeptide chain may be several hundred amino acid units in length. The number of possible arrangements of the different amino acids in a given protein is unbelievably large. In a polypeptide chain made up of 500 amino acid units, the number of possible patterns (given 20 different amino acids) can be expressed by the number 1 followed by 1,100 zeroes. How, then, does a cell form the particular amino acid patterns it requires out of the colossal number of patterns that are possible?

We may presume that the sequence of bases of the DNA molecule is in some way the master pattern, or code, for the sequence of the amino acids in the polypeptide chains of the cellular proteins. In 1954, the British physicist George Gamow suggested that each amino acid is dictated by one sequence of three bases in the DNA molecule. As an example, the sequence cytosine-thymine-thymine (CTT) in the DNA molecule might designate that the amino acid glutamic acid is to be incorporated in the formation of a protein molecule, such as hemoglobin (fig. 3.5). Thus, the DNA code is to be found in triplets— that is, three bases taken together code one amino acid. It should be noted that only one of the two strands of the DNA molecule serves as the genetic code. Biochemical chaos would result if both complementary strands of DNA were to encode information.

The importance of the genetic code cannot be overstated. Let us assume the following sequence of bases in one of the strands of the DNA molecule: . . . CGT ATC GTA AGC . . . , and that the triplets in this sequence specify four amino acids, designated R, I, P, and E. In other words, the following code exists: CGT=R, ATC=I, GTA=P, and AGC=E. This section of the DNA molecule thus specifies RIPE, and the message to make RIPE is passed on by this particular sequence from the nucleus to the cytoplasm of the cell. The message continues to flow out in the living cell, and RIPE will be made, copy after copy.

But what if a chance mishap occurs in one of the triplets? Let us say that T in the second triplet is substituted for A, so that the triplet reads TTC instead of ATC, which would specify A instead of I. The word would now be RAPE instead of RIPE. In a publishing analogy, think of the difference that this continually misprinted word would have throughout a novel. And just as a misprinted word can alter or destroy the meaning of a sentence, so can an altered protein in the body fail to express its intended purpose. Sometimes the error is not tragic, but often the organism is debilitated by the misprint.

As a striking example, we may consider the case of individuals afflicted with sickle-cell anemia. The hemoglobin molecule in sickle-cell anemic patients is biochemically abnormal. In 1956, Vernon Ingram, then at Cambridge

University, succeeded in breaking down hemoglobin, a large protein molecule, into several peptide fragments containing short sequences of identifiable amino acids. Normal hemoglobin and sickle-cell hemoglobin yielded the same array of peptide fragments, with a single exception. In one of the peptide fragments of sickle-cell hemoglobin, the amino acid *glutamic acid* had been replaced at one point in the chain by *valine* (fig. 3.5). The sole difference in chemical composition between normal and sickle-cell hemoglobin is the substitution of only one amino acid unit among several hundred. The detrimental effect of sickle-cell anemia is thus traceable to an exceedingly slight alteration in the structure of the protein molecule. This, in turn, is associated with a highly localized genetic change, or mutation, in one of the base pairs in the DNA molecule in the chromosome (fig. 3.5).

The DNA molecule, like a tape recording, carries specific messages for the synthesis of a wide variety of proteins. However, there are complexities in the process of protein synthesis. For example, the DNA molecule does not directly form protein but works in a complex way through a secondary form of nucleic acid, *ribonucleic acid,* or *RNA*. It is sufficient for our present purpose to recognize that a gene is a coded sequence in the DNA molecule. From a functional point of view, we may say that the gene is *a section of the DNA molecule (about 1,500 base pairs) involved in the determination of the amino acid sequence of a single polypeptide chain of a protein.*

Frequency of Mutations

The rate at which any single gene mutates is generally low, but constant. The average rate of mutation per gene in the fruit fly *(Drosophila melanogaster)* is thought to be about 1 in 100,000 gametes. In other words, any given gene, on the average, mutates approximately once in every 100,000 sperm cells or egg cells produced. Some genes change more often than others. Some indication of the rate of mutation can be obtained from the incidence of a genetic disorder in the population. In humans, a form of dwarfism known as *achondroplasia* occurs in 1 child in nearly 12,000 born to normal parents. Affected infants are small and disproportionate, with abnormally short arms and legs (fig. 3.6). The defect is evident at birth, and most achondroplastic dwarfs fail to survive beyond the first year. Only about 20 percent of the affected individuals reach adulthood. This condition is caused by a dominant mutant gene. Since the parents are normal, each new case of this dominant disorder must result from a newly mutated gene originating in either the sperm or the egg cell. Twelve thousand births represent a total contribution of 24,000 gametes. Accordingly, a new mutation for this peculiar type of dwarfism arises at a rate of 1 in 24,000 gametes, or roughly 4 in 100,000 gametes (4×10^{-5}).

We hasten to point out that the foregoing calculated rate of mutation is undoubtedly too high for a single gene. At the outset, the estimate is unreliable because it is not based on a sufficiently large population sample of the incidence of achondroplasia. In addition, there is evidence that this deformity may result

Figure 3.5 **The abnormal hemoglobin** that occurs in sickle-cell anemia (hemoglobin S) is the consequence of an alteration of a single base of the triplet of DNA that specifies a particular amino acid (glutamic acid) in normal hemoglobin (hemoglobin A). A simple base-pair switch, or mutation, in the DNA molecule results in the replacement of glutamic acid in hemoglobin A by another amino acid, valine, in hemoglobin S. (Based on studies by Vernon M. Ingram.)

Figure 3.6 **Achondroplastic dwarfism,** a dominant genetic disorder in which the affected infant has inherited abnormally short arms and legs. (Courtesy of the Department of Pediatrics, Tulane Medical School.)

from mutations at two or more different gene loci. Hence, the rate of 4×10^{-5} may actually be the interaction of two or three separate rates. Other methods of gauging mutation rates also have noteworthy complications. In our present state of knowledge, the mutation rate in humans for any single gene locus must be considered as a very rough estimate. The consensus is that the mutation rate in humans is in the order of magnitude of one per 100,000 gametes (1×10^{-5}).

Overall Mutation Rate

Although the mutation rate of a single gene may be low, the mutability of the organism as a whole is obviously much higher when we account for the total complement of genes possessed by an individual. A single unfertilized human egg contains an estimated three billion (3×10^{9}) base pairs in its nuclear DNA. This incredible number of nucleotide pairs can carry information corresponding to two million genes. If all these genes were functional, it is likely that the load of mutation would be intolerable to the organism. Indeed, theoretical geneticists have concluded that a reasonable number of functional genes in humans is 50,000. Even this number is overwhelming when one considers the comparatively small number of genetic defects in humans with a simple mode of inheritance. We may presume that not all single-base changes (mutations) have adverse effects or cause overt disease. We also acknowledge that the function of the vast bulk of nuclear DNA in humans remains to be elucidated.

Let us suppose that there are 50,000 human genes in which a mutation (simple base substitution) can lead to a phenotypically noticeable defect. The average rate of mutation per gene in humans is generally held to be about 1 per 100,000 gametes. If the value of $1/100,000$ mutation per gene (1×10^{-5}) is multiplied by the conservative figure of 50,000 genes (5×10^{-4}), then the average number of new mutations per gamete is 0.50. Thus, one of every two gametes produced by a person would bear a new mutant gene capable of having an adverse effect.

Each offspring, of course, is the product of two gametes. Let us envision the random union of sperm cells and egg cells to form zygotes, and say that 50 percent of all gametes bear one newly mutated gene. The chance that an egg cell containing a newly mutated gene will be fertilized by a sperm cell carrying a new mutated gene is 1 in 4, or 25 percent. (Recall that the probability that both coins will show heads when tossed together is $\frac{1}{2} \times \frac{1}{2} = \frac{1}{4}$.) The chance that a zygote will be free of a new mutant gene is also 25 percent (like the probability of two tails). Finally, the chance that a zygote will carry one newly mutated gene contributed by either the sperm cell or the egg cell is 50 percent (similar to the probability of the combination of a head and a tail when two coins are tossed together). It should be apparent that 75 percent of all human offspring would contain at least one newly mutated gene. This assertion, based on crude computations, is in accord with currently accepted notions. As expressed by the English geneticist Harry Harris, every newborn infant, on the average, may be "expected, as a result of a new mutation in either of its parents, to synthesize at least one structurally variant enzyme or protein." Viewed in this manner, the phenomenon of mutation should become very real to us.

It is of interest that of all human pregnancies that continue longer than four or five weeks, 12 to 15 percent end as abortions before the end of the 27th week of pregnancy, and 2 percent terminate as stillbirths. Moreover, the proportion of zygotes lost earlier than the fourth week of pregnancy has been estimated at 10 to 15 percent. Taken all together, 30 percent of all human embryos fail to survive *in utero*. Not all failures, of course, are due to detrimental mutations, but the mutation phenomenon undoubtedly is an important contributing factor.

Harmful Nature of Mutations

Most by far of the gene mutations observable today in organisms are changes for the worse. This is not unexpected. Existing populations of organisms are products of a long evolutionary past. The genes that are now normal to the members of a population represent the most favorable mutations selectively accumulated over eons of time. The chance that a new mutant gene will be more advantageous than an already established favorable gene is slim. Nonetheless, if the environment were to change, the previously adverse mutant gene might prove to be beneficial in the new environmental situation. The microscopic water flea, *Daphnia,* thrives at a temperature of 20°C and cannot survive when the temperature rises to 27°C. A mutant strain of this water flea is known that requires temperatures between 25°C and 30°C and cannot live at 20°C. Thus, at high temperatures, the mutant gene is essential to the survival of the water fleas. This little episode reveals an important point: *A mutation that is inferior in the environment in which it arose may be superior in another environment.* Skeptics might contend that this proverbial declaration has no relevancy to complex higher organisms, including humans. Interestingly, the maxim derives overwhelming support from the varied frequencies of the gene for sickle-cell anemia in human populations. The obviously harmful sickling gene may actually confer an advantage to its carriers in certain geographical localities (see chapter 7).

The process of mutation furnishes the genetic variants that are the raw materials of evolution. Ideally, mutations should arise only when advantageous, and only when needed. This, of course, is fanciful thinking. Mutations occur irrespective of their usefulness or uselessness. The mutations responsible for achondroplasia and retinoblastoma (malignant eye tumors) in humans are certainly not beneficial. But novel inheritable characters repeatedly arise as a consequence of mutation. The harsh price of evolutionary potentialities for a population is the continual occurrence and elimination of mutant genes with detrimental effects. In evolutionary terms, a population must depend on the occasional errors that occur in the copying process of its genetic material if it is to continue to evolve.

Selected Readings

Asimov, I. 1962. *The genetic code.* New York: Signet Science Library.

Crick, F. H. C. 1954. The structure of the hereditary material. *Scientific American,* October, pp. 54–61.

———. 1962. The genetic code. *Scientific American,* October, pp. 66–74.

———. 1966. The genetic code: III. *Scientific American,* October, pp. 55–62.

Crow, J. F. 1959. Ionizing radiation and evolution. *Scientific American,* September, pp. 138–60.

Dobzhansky, T. 1964. *Heredity and the nature of man.* New York: The New American Library.

Ingram, V. 1958. How do genes act? *Scientific American,* January, pp. 68–74.

Kornberg, A. 1968. The synthesis of DNA. *Scientific American,* October, pp. 64–78.

Miller, R. W. 1969. Delayed radiation effects in atomic-bomb survivors. *Science* 166: 569–74.

Muller, H. J. 1955. Radiation and human mutation. *Scientific American,* November, pp. 58–68.

Nirenberg, M. 1963. The genetic code: II. *Scientific American,* March, pp. 80–94.

Papazian, H. P. 1967. *Modern genetics.* New York: W. W. Norton & Co.

Schull, W. J., and Neel, J. V. 1958. Radiation and the sex ratio in man. *Science* 128: 343–48.

Sinsheimer, R. 1962. Single-stranded DNA. *Scientific American,* July, pp. 109–16.

Spiess, E. B. 1977. *Genes in populations.* New York: John C. Wiley & Sons.

United Nations Scientific Committee. 1967. *The genetic risk from radiation.* New York: United Nations.

Wallace, B. 1966. *Chromosomes, giant molecules, and evolution.* New York: W. W. Norton & Co.

Wallace, B., and Dobzhansky, T. 1959. *Radiation, genes, and man.* New York: Holt, Rinehart & Winston.

Watson, J. D. 1968. *The double helix.* New York: Atheneum Publishers.

4
Genetic Equilibrium

The opening chapter introduced us to a natural population of bullfrogs that conspicuously contained at one time several hundred multilegged variants. We surmised that the multilegged anomaly was an inherited condition, transmitted by a detrimental recessive gene. The multilegged frogs disappeared in nature as dramatically as they appeared. They unquestionably failed to reproduce and leave descendants. Now, let us imagine that the multilegged frogs were as reproductively fit as their normal kin. Would the multilegged trait eventually still be eliminated from the population?

A comparable question was posed to the English geneticist R. C. Punnett at the turn of the century. He was asked to explain the prevalence of blue eyes in humans in view of the acknowledged fact that the blue-eyed condition was a recessive characteristic. Would it not be the case that the dominant brown-eyes trait would in time supplant the blue-eyed state in the human population? The answer was not self-evident, and Punnett sought out his colleague Godfrey H. Hardy, the astute mathematician at Cambridge University. Hardy had only a passing interest in genetics, but the problem intrigued him as a mathematical one. The solution, which we shall consider below, ranks as one of the most fundamental laws of genetics and evolution. As fate has it, Hardy's formula was arrived at independently in the same year (1908) by a physician, Wilhelm Weinberg, and the well-known equation presently bears both their names.

Mendelian Inheritance

We may recall that the genetic constitutions, or genotypes, of the normal and multilegged frogs have been designated as AA (normal), Aa (normal but a carrier), and aa (multilegged). The kinds and proportions of offspring that can arise from matings involving the three genotypes are illustrated in figure 4.1. Six different types of matings are possible. The mating $AA \times AA$ gives

Figure 4.1 Six possible mating types with respect to one pair of genes, and the kinds and percentages of offspring from each type of mating. Normal frogs are either homozygous dominant (*AA*) or heterozygous (*Aa*); multilegged frogs are recessive (*aa*). The sex of the parent is not denoted; in crosses of unlike genotypes (such as *AA* × *Aa*), either genotype may be the male or the female.

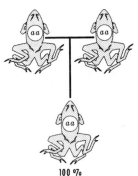

rise solely to normal homozygous offspring, *AA*. Two kinds of progeny, *AA* and *Aa* in equal proportions, result from the cross of a homozygous normal parent (*AA*) and a heterozygous parent (*Aa*). The mating of two heterozygotes (*Aa* × *Aa*) produces *AA*, *Aa*, and *aa* offspring in the classical Mendelian ratio of 1:2:1. Only heterozygous offspring (*Aa*) emerge from the mating *AA* × *aa*. Both heterozygous (*Aa*) and recessive (*aa*) progeny, in equal numbers, arise from the cross of a heterozygous parent (*Aa*) and a recessive parent (*aa*). Lastly, two recessive parents (*aa* × *aa*) produce only recessive offspring (*aa*).

These principles of Mendelian inheritance merely inform us that certain kinds of offspring can be expected from certain types of matings. If we are interested in following the course of a population from one generation to the next, then additional factors enter the scene.

Random Mating

One important factor that influences the genetic composition of a population is the system of mating among individuals. The simplest scheme of breeding activity in a population is called *random mating,* wherein any one individual has an equal chance of pairing with any other individual. Random mating does not mean promiscuity; it simply means that those who choose each other as mating partners do not do so on the basis of similarity or dissimilarity in a given trait or gene.

The absence of preferential mating in a population has interesting consequences. Let us suppose that random mating prevails in our particular population of bullfrogs. This assemblage of frogs is ordinarily very large, numbering several thousand individuals. For ease of presentation, however, the size of the population is reduced to 48 males and 48 females. Moreover, for each sex, we may simplify the mathematical computations by assuming that 36 are normal (12 *AA* and 24 *Aa*) and 12 are multilegged (*aa*). Accordingly, one-quarter of the individuals of each sex are homozygous dominant, one-half are heterozygous, and one-quarter are recessive. Now, if mating occurs at random, will the incidence of multilegged frogs decrease, increase, or remain the same in the next generation?

The problem may be approached by determining how often a given type of mating occurs. Here we will use the multiplication rule of probability: *The chance that two independent events will occur together is the product of their chances of occurring separately.* The proportion of *AA* males in our arbitrary bullfrog population is 1/4. We may also say that the chance that a male bullfrog is *AA* is 1/4. Likewise the probability that a female bullfrog is *AA* is 1/4. Consequently, the chance that an *AA* male will "occur together," or mate, with an *AA* female is 1/16 (1/4 × 1/4). The computations for all types of matings can be facilitated by coupling the males and females in a multiplication table, as shown in table 4.1.

Table 4.1
Random Mating

Female (♀)	Male (♂)		
	1/4 *AA*	2/4 *Aa*	1/4*aa*
1/4 *AA*	1/16 *AA* × *AA*	2/16 *AA* × *Aa*	1/16 *AA* × *aa*
2/4 *Aa*	2/16 *Aa* × *AA*	4/16 *Aa* × *Aa*	2/16 *Aa* × *aa*
1/4 *aa*	1/16 *aa* × *AA*	2/16 *aa* × *Aa*	1/16 *aa* × *aa*

Table 4.1 shows that there are nine combinations of mated pairs, and that some types occur more frequently than others. It may be helpful to express the frequencies in terms of actual numbers. Thus, for a total of 48 matings, 3 (= 1/16 × 48) would be *AA* ♀ × *AA* ♂, 6 (= 2/16 × 48) would be *AA* ♀ × *Aa* ♂, 12 (= 4/16 × 48) would be *Aa* ♀ × *Aa* ♂, and so forth. The numbers of each type of mating are listed in table 4.2.

Table 4.2
First Generation of Offspring

Type of Mating (Female x Male)	Number of Each Type of Mating*	Number of Offspring		
		AA	*Aa*	*aa*
AA × *AA*	3	12		
AA × *Aa*	6	12	12	
AA × *aa*	3		12	
Aa × *AA*	6	12	12	
Aa × *Aa*	12	12	24	12
Aa × *aa*	6		12	12
aa × *AA*	3		12	
aa × *Aa*	6		12	12
aa × *aa*	3			12
		48	96	48
		(25%)	(50%)	(25%)

*Based on a total of 48 matings.

Our next step is to ascertain the kinds and proportions of offspring from each mating. We shall assume that each mated pair yields the same number of offspring—for simplicity, four offspring. (This is an inordinately small number, as a single female bullfrog can deposit well over 10,000 eggs.) We also take for granted that the genotypes of the four progeny from each mating are those that are theoretically possible in Mendelian inheritance (see fig. 4.1). For example, if the parents are *Aa* × *Aa,* their offspring will be 1 *AA,* 2 *Aa,* and 1 *aa.* In another instance, if the parents are *AA* × *Aa,* then the offspring will be 2 *AA* and 2 *Aa.* The outcome of all crosses is shown in table 4.2. It is important to note that the actual numbers of offspring recorded in table 4.2 are related to the frequencies of the different types of matings. For example, the mating of an *AA* female with an *Aa* male occurs six times; hence, the numbers of offspring are increased sixfold (from 2 each of *AA* and *Aa* to 12 each of the two genotypes).

An examination of table 4.2 reveals that the kinds and proportions of individuals in the new generation of offspring are exactly the same as in the parental generation. There has been no change in the ratio of normal frogs (75 percent *AA* and *Aa*) to multilegged frogs (25 percent *aa*). In fact, the proportions of phenotypes (and genotypes) will remain the same in all successive generations, provided that the system of random mating is continued.

Gene Frequencies

There is a less tedious method of arriving at the same conclusion. Rather than figure out all the matings that can possibly occur, we need only to consider the genes that are transmitted by the eggs and sperm of the parents. Let us assume that each parent produces only 10 gametes. The 12 homozygous dominant males (*AA*) of our arbitrary initial population can contribute 120 sperm cells to the next generation, each sperm containing one *A.* The 24 heterozygous males (*Aa*) can transmit 240 gametes, 120 of them with *A* and 120 with *a.* The remaining 12 recessive males (*aa*) can furnish 120 gametes, each with *a.* The total pool of genes provided by all males will be 240 *A* and 240 *a,* or 50 percent of each kind. Expressed as a decimal fraction, the frequency of gene *A* is 0.5; of *a,* 0.5.

Since the females in our population have the same genetic constitutions as the males, their gametic contribution to the next generation will also be 0.5 *A* and 0.5 *a.* The eggs and sperm can now be united at random in a genetical checkerboard (fig. 4.2).

It should be evident from figure 4.2 that the distribution of genotypes among the offspring is 0.25 *AA* : 0.50 *Aa* : 0.25 *aa.* The random union of eggs and sperm yields the same result as the random mating of parents (refer to table 4.2). Thus, using two different approaches, we have answered the question posed in the introductory remarks to this chapter. *If the multilegged frogs are equally as fertile as the normal frogs and leave equal numbers of offspring each generation, then these anomalous frogs will persist in the population with the same frequency from one generation to the next.*

Figure 4.2   **Random union** of eggs and sperm yields the same outcome as the random mating of parents (refer to table 4.2).

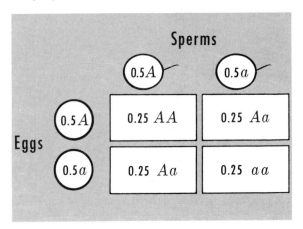

Hardy-Weinberg Equilibrium

A population in which the proportions of genotypes remain unchanged from generation to generation is in *equilibrium*. The fact that a system of random mating leads to a condition of equilibrium was uncovered independently by G. H. Hardy and W. Weinberg, and has come to be widely known as the *Hardy-Weinberg Equilibrium*. This theorem states that the proportions of *AA, Aa,* and *aa* genotypes, as well as the proportions of *A* and *a* genes, will remain constant from generation to generation, provided that the bearers of the three genotypes have equal opportunities of producing offspring in a large, randomly mating population.

The above statement can be translated into a simple mathematical expression. If we let p be the frequency of the gene *A* in the population, and q equal the frequency of its allele, *a,* then the distribution of the genotypes in the next generation will be $p^2 AA : 2pq \ Aa : q^2 aa$. This relationship may be verified by the use, once again, of a genetical checkerboard (fig. 4.3). Mathematically inclined readers will recognize that $p^2 : 2pq : q^2$ is the algebraic expansion of the binomial $(p + q)^2$. The frequencies of the three genotypes (0.25 *AA* : 0.50 *Aa* : 0.25 *aa*) in our bullfrog population under the system of random mating is the expanded binomial $(0.5 + 0.5)^2$.

We may consider another arbitrary population in the equilibrium state. Suppose that the population consists of 16 *AA*, 48 *Aa*, and 36 *aa* individuals. We may assume, as before, that 10 gametes are contributed by each individual to the next generation. All the gametes transmitted by the 16 *AA* parents (numerically, 160) will contain the *A* gene, and half the gametes (240) provided by the *Aa* parents will bear the *A* gene. Thus, of the 1,000 total gametes in the population, 400 will carry the *A* gene. Accordingly, the frequency of

Figure 4.3   **The distribution of genotypes** in the next generation is p^2 AA : $2pq$ Aa : q^2aa (Hardy-Weinberg formula).

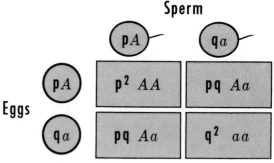

gene A is 0.4 (designated p). In like manner, it can be shown that the frequency of gene a is 0.6 (q). Substituting the numerical values for p and q in the Hardy-Weinberg formula, we have:

p^2 AA : $2pq$ Aa : q^2 aa
$(0.4)^2$ AA : $2(0.4)$ (0.6) Aa : $(0.6)^2$ aa
0.16 AA : 0.48 Aa : 0.36 aa.

Hence, the proportions of the three genotypes are the same as those of the preceding generation.

It should also be clear that a recessive trait, such as blue eyes in humans, will not become rare just because it is governed by a recessive gene. Nor can the dominant brown-eyed condition become widespread simply by virtue of its dominance. Whether a given gene is common or rare is controlled by other factors, particularly natural selection.

Estimating the Frequency of Heterozygotes

It comes as a surprise to many readers to discover that the heterozygotes of a rare recessive abnormality are rather common instead of being comparatively rare. Recessive albinism may be used as an illustration. The frequency of albinos is about 1/20,000 in human populations. When the frequency of the homozygous recessive (q^2) is known, the frequency of the recessive gene (q) can be calculated, as follows:

$q^2 = 1/20,000 = 0.00005$

$q = \sqrt{0.00005} = 0.007$
$= $ about 1/140 (frequency of recessive gene).

The heterozygotes are represented by $2pq$ in the Hardy-Weinberg formula. Accordingly, the frequency of heterozygous carriers of albinism can be calculated as follows:

$$q = 0.007$$

$$p = 1 - 0.007 = 0.993$$

$$\therefore 2pq = 2\ (0.993 \times 0.007) = 0.014$$

$$= \text{about } 1/70 \text{ (frequency of heterozygote).}$$

Thus, although 1 person in 20,000 is an albino (recessive homozygote), about 1 person in 70 is a heterozygous carrier. There are 280 times as many carriers as affected individuals! It bears emphasizing that the rarity of a recessive disorder does not signify a comparable rarity of heterozygous carriers. In fact, when the frequency of the recessive gene is extremely low, nearly all the recessive genes are in the heterozygous state.

Implications

The Hardy-Weinberg theorem is entirely theoretical. The set of underlying assumptions can scarcely be fulfilled in any natural population. We implicitly assume the absence of recurring mutations, the absence of any degree of preferential matings, the absence of differential mortality or fertility, the absence of immigration or emigration of individuals, and the absence of fluctuations in gene frequencies due to sheer chance. But therein lies the significance of the Hardy-Weinberg theorem. In revealing the conditions under which evolutionary change cannot occur, it brings to light the possible forces that could cause a change in the genetic composition of a population. The Hardy-Weinberg theorem thus depicts a static situation.

An understanding of Hardy-Weinberg equilibrium provides a basis for recognizing the forces that permit evolutionary change. A few of the more obvious factors that prevent a natural population from attaining stationary equilibrium are: (1) mutation, (2) natural selection, (3) chance events in a small population (genetic drift), and (4) migration. These factors or forces may profoundly modify the gene frequencies in natural populations. In the next chapter, we shall consider how one of the forces, natural selection, can lead to dynamic evolutionary changes.

Selected Readings

Bodmer, W. F., and Cavalli-Sforza, L. L. 1976. *Genetics, evolution, and man.* San Francisco: W. H. Freeman and Co.

Dobzhansky, T. 1970. *Genetics of the evolutionary process.* New York: Columbia University Press.

Dunn, L. C., and Dobzhansky, T. 1952. *Heredity, race, and society.* New York: The New American Library.

Hardy, G. H. 1908. Mendelian proportions in a mixed population. *Science* 28: 49–50.

Li, C. C. 1976. *First course in population genetics.* Pacific Grove, Calif.: Boxwood Press.

Mettler, L. E., and Gregg, T. G. 1969. *Population genetics and evolution.* Englewood Cliffs, N.J.: Prentice-Hall.

Spiess, E. B. 1977. *Genes in populations.* New York: John C. Wiley & Sons.

Stern, C. 1943. The Hardy-Weinberg law. *Science* 97: 137–38.

———. 1973. *Principles of human genetics.* San Francisco: W. H. Freeman and Co.

Wallace, B. 1968. *Topics in population genetics.* New York: W. W. Norton & Co.

5
Concept of Selection

We have already remarked that the multilegged anomaly, which appeared in 1958 in a local bullfrog population, has not been detected since that occasion. The supposition was made that the multilegged trait was governed by a recessive mutant gene. Some of us might presume that the mutant gene responsible for the abnormality has disappeared entirely from the population. But can a detrimental mutant gene be completely eradicated from natural populations of organisms, even in the face of the severest form of selection? Most persons are frankly puzzled when they are informed that the answer is *"No."* Yet, our knowledge of the properties of mutation and selection expressly permits a firm negative reply.

Selection against Recessive Defects

In our consideration of the Hardy-Weinberg Equilibrium in the preceding chapter, we assumed that the mutant multilegged frogs (*aa*) were as reproductively fit as their normal kin (*AA* and *Aa*) and left equal numbers of living offspring each generation. Now, however, let us presume that all multilegged individuals fail to reach sexual maturity generation after generation. Will the incidence of the multilegged trait decline to a vanishing point?

We may start with the same distribution of individuals in the initial generation previously postulated in chapter 4, namely, 24 *AA*, 48 *Aa*, and 24 *aa*, with the sexes equally represented. Since the multilegged frogs (*aa*) are unable to participate in breeding, the parents of the next generation comprise only the 24 *AA* and 48 *Aa* individuals. The heterozygous types are twice as numerous as the homozygous dominants; accordingly, two-thirds of the total breeding members of the population are *Aa* and one-third are *AA*. We may once again employ a genetical checkerboard (table 5.1) to ascertain the different types of matings and their relative frequencies.

Table 5.1
Matings and Relative Frequencies

Female	Male ($\male$)	
($\female$)	1/3 *AA*	2/3 *Aa*
1/3 *AA*	1/9 *AA* × *AA*	2/9 *AA* × *Aa*
2/3 *Aa*	2/9 *Aa* × *AA*	4/9 *Aa* × *Aa*

The frequencies of the different matings shown in table 5.1 may be expressed as whole numbers rather than fractions. Given a total of 36 matings, 4 (= 1/9 × 36) would be *AA* $\female$ × *AA* $\male$, 8 (= 2/9 × 36) would be *AA* $\female$ × *Aa* $\male$, 8 (= 2/9 × 36) would be *Aa* $\female$ × *AA* $\male$, and 16 (=4/9 × 36) would be *Aa* $\female$ × *Aa* $\male$. These numbers are recorded in table 5.2.

Table 5.2
First Generation Offspring after Complete Selection

Type of Mating (Female × Male)	Number of Each Type of Mating*	Number of Offspring		
		AA	*Aa*	*aa*
AA × *AA*	4	16		
AA × *Aa*	8	16	16	
Aa × *AA*	8	16	16	
Aa × *Aa*	16	16	32	16
		64	64	16
		(44.44%)	(44.44%)	(11.11%)

*Based on a total of 36 matings.

Our next task is to determine the outcome of each type of cross. We shall assume that each mated pair contributes an equal number of progeny to the next generation (say, four offspring). As revealed in table 5.2, the offspring are distributed according to Mendelian ratios, and the actual numbers of offspring reflect the frequencies of the different kinds of matings. For example, a single *AA* $\female$ × *Aa* $\male$ mating yields 4 offspring in the Mendelian ratio of 2 *AA* : 2 *Aa*. There are, however, eight matings of this kind; the numbers of offspring are correspondingly increased to 16 *AA* and 16 *Aa*.

Even though all the multilegged frogs fail to reproduce, the detrimental recessive genes are still transmitted to the first generation. The emergence of

Table 5.3
Effects of Complete Selection against a Recessive Trait

Generations	Gene Frequency	Recessive Homozygotes %	Heterozygotes %	Dominant Homozygotes %
0	0.500	25.00	50.00	25.00
1	0.333	11.11	44.44	44.44
2	0.250	6.25	37.50	56.25
3	0.200	4.00	32.00	64.00
4	0.167	2.78	27.78	69.44
8	0.100	1.00	18.00	81.00
10	0.083	0.69	15.28	84.03
20	0.045	0.20	8.68	91.12
30	0.031	0.10	6.05	93.85
40	0.024	0.06	4.64	95.30
50	0.020	0.04	3.77	96.19
100	0.010	0.01	1.94	98.05

multilegged frogs in the first generation stems from the matings of two heterozygous frogs. However, as seen from table 5.2, the frequency of the multilegged trait (*aa*) decreases from 25 percent to 11.11 percent in a single generation.

The effects of complete selection against the multilegged frogs in subsequent generations can be determined by the foregoing method of calculation, but the lengthy tabulations can be wearisome. At this point we may apply a formula that will establish in a few steps the frequency of the recessive gene after any number of generations of complete selection:

$$q_n = \frac{q_0}{1 + nq_0}.$$

In the above expression, q_0 represents the initial or original frequency of the recessive gene, and q_n is the frequency after n generations. Thus, with the initial value of $q_0 = 0.5$, the frequency of the recessive gene after two generations ($n = 2$) will be:

$$q_2 = \frac{q_0}{1 + 2q_0} = \frac{0.5}{1 + 2(0.5)} = \frac{0.5}{2.0} = 0.25.$$

If the frequency of the recessive gene itself (*a*) is q, then the frequency of the recessive individual (*aa*) is q^2. Accordingly, the frequency of the recessive homozygote is $(0.25)^2$, or 0.0625 (6.25 percent). In the second generation, therefore, the incidence of the multilegged trait drops to 6.25 percent.

If we perform comparable calculations through several generations, we emerge with a comprehensive picture that is tabulated in table 5.3 and portrayed in figure 5.1. In the third generation, the frequency of the recessive

Figure 5.1 Effects of complete selection against recessive homozygotes (*aa*) occurring initially ("0" generation) at a frequency of 25 percent. The effectiveness of selection in reducing the incidence of the recessive trait decreases with successive generations. The frequency of recessive homozygotes drops markedly from 25 percent to 6.25 percent in two generations. However, 8 generations are required to reduce the incidence of the recessive trait to 1.0 percent, 30 generations are needed to achieve a reduction to 0.1 percent, and approximately 100 generations to depress the frequency to 0.01 percent.

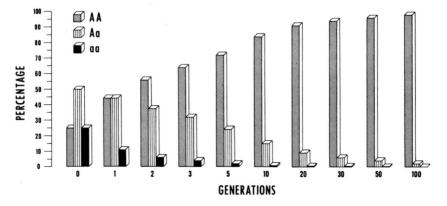

homozygote declines to 4.0 percent. Progress in the elimination of the multilegged trait is initially rapid, but becomes slower as selection is continued over many successive generations. About 20 generations are required to depress the incidence of the multilegged trait to 2 in 1,000 individuals (0.20 percent). Ten additional generations are necessary to effect a reduction to 1 in 1,000 individuals (0.10 percent). Thus, as a recessive trait becomes rarer, selection against it becomes less effective. The reason is quite simple: Only very few recessive homozygotes are exposed to the action of selection. The now rare recessive gene (*a*) is carried mainly by heterozygous individuals (*Aa*), where it is sheltered from selection by its normal dominant partner (*A*).

Significance of the Heterozygote

When the frequency of an abnormal recessive gene becomes very low, most affected offspring (*aa*) will come from matings of two heterozygous carriers (*Aa*). For example, in the human population, the vast majority of newly arising albino individuals (*aa*) in a given generation (more than 99 percent of them) will come from normally pigmented heterozygous parents. Considerations of this kind led us to postulate in chapter 1 that the multilegged frogs that appeared suddenly in the natural population were derived from normal-legged heterozygous parents (refer to fig. 1.4).

Detrimental recessive genes in a population are unquestionably harbored mostly in the heterozygous state. As shown in table 5.4, the frequency of

Table 5.4
Frequencies of Recessive Homozygotes and Heterozygous Carriers

Frequency of Homozygotes (*aa*)	Frequency of Heterozygous Carriers (*Aa*)	Ratio of Carriers to Homozygotes
1 in 500 (Sickle-Cell Anemia)[a]	1 in 10	50:1
1 in 1,000 (Cystic Fibrosis)	1 in 16	60:1
1 in 6,000 (Tay-Sachs Disease)[b]	1 in 40	150:1
1 in 20,000 (Albinism)	1 in 70	285:1
1 in 25,000 (Phenylketonuria)	1 in 80	310:1
1 in 50,000 (Acatalasia)[c]	1 in 110	460:1
1 in 1,000,000 (Alkaptonuria)	1 in 500	2,000:1

[a]Based on incidence among American blacks.
[b]In the United States, the disease occurs once in 6,000 Jewish births and once in 500,000 non-Jewish births.
[c]Based on prevalence rate among the Japanese.

heterozygous carriers is many times greater than the frequency of homozygous individuals afflicted with the trait. Thus, an extremely rare disorder, like alkaptonuria (blackening of urine), occurs in 1 in one million persons. This detrimental gene, however, is carried in the hidden state by 1 out of 500 persons. There are two thousand as many genetic carriers of alkaptonuria as there are individuals afflicted with this defect. For another recessive trait, cystic fibrosis, 1 out of 1,000 individuals is affected with this homozygous trait. One of 16 persons is a carrier of cystic fibrosis. In modern genetic counseling programs, an important consideration has been the development of simple, inexpensive means of detecting heterozygous carriers of inherited disorders.

Interplay of Mutation and Selection

Theoretically, if the process of complete selection against the recessive homozygote were to continue for several hundred more generations, then the abnormal recessive gene would be completely eliminated, and the population would consist uniformly of normal homozygotes (*AA*). But, *in reality*, the steadily diminishing supply of abnormal recessive genes is continually being replenished by recurrent mutations from normal (*A*) to abnormal (*a*). Mutations from *A* to *a*, which inevitably occur from time to time, were not taken into account in our foregoing determinations. Mutations, of course, cannot be ignored.

All genes undergo mutation at some definable rate. If a certain proportion of *A* genes are converted into *a* alleles in each generation, then the population will at all times carry a certain amount of the recessive mutant gene (*a*) despite selection against it. Without any sophisticated calculations, it can be shown that a point will be reached at which the number of the abnormal recessive genes eliminated by selection just balances the number of the same recessive gene produced by mutation. An analogy shown in figure 5.2 A will help in visualizing this circumstance. The water level in the beaker remains constant when the rate at which water enters the opening of the beaker equals the rate at which it leaves the hole in the side of the beaker. In other words, a state of equilibrium is reached when the rate at which the recessive gene is replenished by mutation equals the rate at which it is lost by selection. It

---

Figure 5.2 **Interplay of detrimental mutant genes** (water from faucet) and their elimination by selection (water escaping through hole) in a population (beaker) containing a pool of the harmful genes (water in beaker). *A.* State of genetic equilibrium (constant water level in beaker) when the rates at which water enters and leaves the beaker are equal. *B.* Effect on an increase in mutation rate (increased flow of faucet water) as might be expected from the continued widespread use of ionizing radiation. A new equilibrium (new constant water level) is established, but the frequency of the detrimental gene in the population is higher (higher water level in beaker). *C.* Effect of reducing selection pressure (decreased exit of water) as a consequence of improving the reproductive fitness of genetically defective individuals by modern medical practices. Mutation rate (inflow of water) is the same as in *A.* The inevitable result is a greater incidence of the harmful mutant gene in the population (higher level of water in the beaker).

A   **B**   **C**

should be clear that it is not mutation alone that governs the incidence of deleterious recessives in a population. The generally low frequency of harmful recessive genes stems from the dual action of mutation and selection. The mutation process tends to increase the number of detrimental recessives; the selection mechanism is the counteracting agent.

What would be the consequences of an increase in the mutation rate? Man today lives in an environment in which high-energy radiation promotes a higher incidence of mutations. We may return to our analogy (fig. 5.2 B). The increased rate of mutation may be envisioned as an increased input of water. The water level in the beaker will rise and water will escape more rapidly through the hole in the side of the beaker. Similarly, mutant genes will be found more frequently in a population, and they will be eliminated at a faster rate from the population. As before, a balance will be restored eventually between mutation and selection, but now the population has a larger store of deleterious genes and a larger number of defective individuals arising each generation.

The supply of defective genes in the human population has already increased through the greater medical control of recessive disorders. The outstanding advances in modern medicine have served to prolong the lives of genetically defective individuals who might otherwise not have survived to reproductive age. This may be compared to partially plugging the hole in the side of the beaker (fig. 5.2 C). The water level in the beaker will obviously rise, as will the amount of deleterious genes in a population. Evidently, the price of our humanitarian principles is the enlargement of our pool of defective genes.

Partial Selection

We have treated above the severest form of selection against recessive individuals. Complete, or 100 percent, selection against a recessive homozygote is often termed *lethal* selection, and the mutant gene is designated as a *lethal* gene. A lethal gene does not necessarily result in the death of the individual, but does effectively prevent him from reproducing or leaving offspring. Not all mutant genes are lethal; in fact, the majority of them have less drastic effects on viability or fertility. A mildly handicapped recessive homozygote may reproduce but may be inferior in fertility to the normal individual. When the reproductive capacity of the recessive homozygote is only half as great as the normal type, he is said to be *semisterile*, and the mutant gene is classified as *semilethal*. A *subvital* recessive gene is one which, in double dose, impairs an individual to the extent that his reproductive fitness is less than 100 percent but more than 50 percent of normal proficiency.

The action of selection varies correspondingly with the degree of detrimental effect of the recessive gene. Figure 5.3 shows the results of different intensities of selection in a population that initially contains 1.0 percent recessive homozygotes. With complete (lethal) selection, a reduction in the

Figure 5.3   Different intensities of selection against recessive homozygotes occurring initially ("0" generation) at a frequency of 1.0 percent. The elimination of recessive individuals per generation proceeds at a slower pace as the strength of selection decreases.

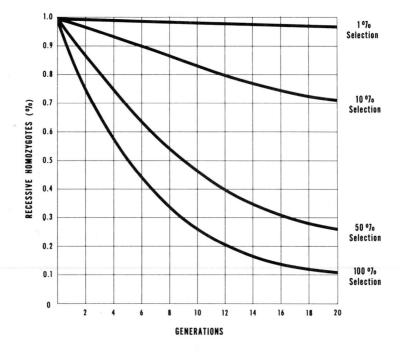

incidence of the recessive trait from 1.0 percent to 0.25 percent is accomplished in 10 generations. Twenty generations of complete selection reduces the incidence to 0.11 percent. When the recessive gene is semilethal (50 percent selection), 20 generations, or twice as many generations as under complete selection, are required to depress the frequency of the recessive homozygote to about 0.25 percent. Selection against a subvital gene (for example, 10 percent selection) results in a considerably slower rate of elimination of the recessive homozygotes. When the homozygote is at a very slight reproductive disadvantage (1.0 percent selection), only a small decline of 0.03 percent (from 1.0 percent to 0.97 percent) occurs after twenty generations. It is evident that mildly harmful recessive genes may remain in a population for a long time.

Selection against Dominant Defects

If complete selection so acts against an abnormal trait caused by a dominant gene (*A*) that none of the *AA* or *Aa* individuals leave any progeny, then all the *A* genes are at once eliminated. In the absence of recurrent mutation, all subsequent generations will consist exclusively of homozygous recessives (*aa*).

Table 5.5

Equilibrium Frequency of a Dominant Gene

Conditions: 1. Size of Population: 500,000 individuals
            2. Mutation Rate ($a \longrightarrow A$): 1 in 100,000
            3. Selection Coefficient (s) : 0.5 (semilethal)

Generation	Recessive Gene a	Dominant Gene A		Newly Mutated	Total
		Left over from Former Generations			
0	,000,000	—		—	—
1	1,000,000*	—		10	10
2	1,000,000	5		10	15
3	1,000,000	5 + 2.5		10	17.5
4	1,000,000	5 + 2.5 + 1.25		10	18.75
5	1,000,000	5 + 2.5 + 1.25 + 0.625		10	19.375
∞	1,000,000	5 + 2.5 + 1.25 + 0.625 + 0.3125 + ...		10	20.00

*The total number of recessive genes should be reduced by the total number of dominant genes each generation, but this minor correction would be inconsequential.

However, we must contend again with ever-occurring mutations and the effects of partial selection. The geneticist Curt Stern, of the University of California, provides us with a simple, clear model of this situation. Imagine a population of 500,000 individuals, all of whom are initially homozygous recessive (aa). Thus, no detrimental dominant genes (A) are present and the population as a whole contains 1 million recessive genes (a). In the first generation, 10 dominant mutant genes arise, as a result of the recessive gene's mutating to the dominant state at a rate of 1 in 100,000 genes. We shall now assume that the dominant mutant gene is semilethal; in other words, only 5 of the newly arisen dominant genes are transmitted to the next, or second, generation. For ease of discussion, this is pictorially shown in figure 5.4 and represented also in table 5.5. It can be seen that the second generation would contain a total of 15 dominant genes—5 brought forth from the first generation and 10 new ones added by mutation. In the third generation, the 5 dominant genes carried over from the first generation would be reduced to 2.5, the 10 dominant alleles of the second generation would be depressed to 5, and 10 new abnormal alleles would arise anew by mutation. The total number of dominant genes would increase slightly each subsequent generation, until a point is reached (about 12 generations) where the rate of elimination of the abnormal dominant gene balances the rate of mutation. In other words, the inflow of new dominant alleles by mutation each generation is balanced by the outflow or elimination of the dominant genes each generation by selection.

The equilibrium frequency of the detrimental dominant gene in a population can be altered by changing the rate of loss of the gene in question. In humans, retinoblastoma, or cancer of the eye in newborn babies, has until

Figure 5.4   Establishment of a constant level of a semilethal dominant gene (*A*) in a population over the course of several generations. The fixed number of new dominant genes introduced each generation through mutations from *a* to *A* eventually exactly balances the number of dominant genes selectively eliminated each generation. In this particular case, an equilibrium is reached (after about 12 generations) when the total number of dominant genes is approximately 20.

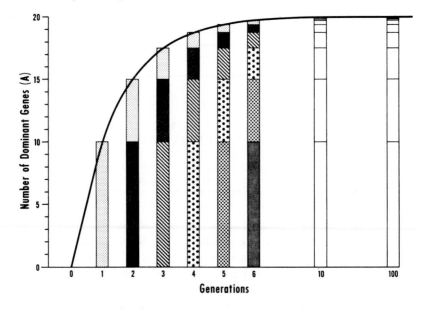

recently been a fatal condition by a dominant mutant gene. With modern medical treatment in developed nations, approximately 60 percent of the afflicted individuals can be saved. The effect of increasing the reproductive fitness of the survivors is to raise the frequency of the abnormal dominant gene in the human population. If the selection coefficient changes from 1.0 to 0.40, a new equilibrium value for the dominant gene for retinoblastoma will be reached after only five to six generations after successful medical therapy has been introduced. Several generations from now, we may expect that inherited retinoblastoma will be approximately three times more common. In other words, cases of retinoblastoma in the newborn will increase from the current frequency of 1 per 40,000 to 1 per 15,000. The accumulation of detrimental genes in the human gene pool is a matter of growing awareness and concern.

Concealed Variability in Natural Populations

From what we have already learned, we should expect to find in natural populations a large number of deleterious recessive genes concealed in the heterozygous state. It may seem that this expectation is based more on theoretical deduction than on actual demonstration. This is not entirely the case.

Penetrating studies by a number of investigators of several species of the fruit fly *Drosophila* have unmistakably indicated an enormous store of recessive mutant genes harbored by individuals in nature. We may take as an illustrative example the kinds and incidence of recessive genes detected in *Drosophila pseudoobscura* from California populations. The following data are derived from the studies of the late geneticist Theodosius Dobzhansky, then at Columbia University but later at the University of California in Davis.

Flies were collected from nature, and a series of elaborate crosses were performed in the laboratory to yield offspring in which one pair of chromosomes carried an identical set of genes. The formerly hidden recessive genes in a given pair of chromosomes were thus all exposed in the homozygous state. All kinds of recessive genes were uncovered in different chromosomes, as exemplified by those unmasked in one particular chromosome, known simply as "the second." About 33 percent of the second chromosomes harbored one or more recessive genes that proved to be lethal or semilethal to flies carrying the second chromosome in duplicate. An astonishing number of second chromosomes—93 percent of them—contained genes that produced subvital or mildly incapacitating effects when present in the homozygous condition. Other unmasked recessive genes resulted in sterility of the flies or severely retarded the developmental rates of the flies. All these flies were normal in appearance when originally taken from nature. It is apparent that very few, if any, outwardly normal flies in natural populations are free of hidden detrimental recessive genes.

Genetic Load in Human Populations

The study of the concealed variability, or genetic load, in humans cannot be approached, for obvious reasons, by the experimental breeding techniques used with fruit flies. Estimates of the genetic load in the human population have been based principally on the incidence of defective offspring from marriages of close relatives (consanguineous marriages). It can be safely stated that every human individual contains at least one newly mutated gene. It can also be accepted that any crop of gametes contains, in addition to one or more mutations of recent origin, at least 10 mutant genes that arose in the individuals of preceding generations and which have accumulated in the population. The average person is said to harbor four concealed lethal genes, each of which, if homozygous, is capable of causing death between birth and maturity. The most conservative estimates place the incidence of deformities to detrimental mutant genes in the vicinity of 2 per 1,000 births. *It is evident that humans are not uniquely exempt from their share of defective genes.*

Muller's AID Plan

Modern medicine, by finding ways to keep alive individuals who carry deleterious genes, encourages the survival of those who more than likely will pass on their defects to future generations. The conventional ethics of medicine are shaken when, with increasing knowledge, it becomes clear that saving the life of a child with a hereditary disorder also ensures the retention and increase of detrimental genes that natural selection ordinarily keeps at very low frequencies. Can we continue indefinitely to load our population with hereditary disabilities?

Some geneticists have looked upon this situation with grave concern. In particular, the late Nobel laureate Hermann J. Muller was most distressed about the continual pollution of the human gene pool. Muller had been predicting genetic disaster since 1935, and throughout his career he was a persuasive and articulate prophet of doom. Muller presented the most vivid portrayal of the impending genetic disintegration of the human species. In the not too distant future, according to Muller, the task of taking care of genetically defective individuals will consume all the energy that society can mobilize. Most everyone will be an invalid. Muller's gloomy forecast was that the human species would end up with two types of individuals: one kind would be so genetically incapacitated as to be wheelchair patients, and the other kind would be somewhat less disabled, but would spend all their time taking care of the first kind. It is unreasonable, Muller contended, to expect medicine to keep up with the problem, especially because medical people themselves in that near or distant future will be subject to the same genetic decomposition. Eventually even the most sophisticated techniques available could no longer suffice to save people from their genetic deterioration. According to Muller, then, humanity is doomed unless positive steps are taken to regulate its genetic endowment.

Muller was convinced that the human species would be unable to reduce the load of unfavorable genes. He suggested as a countermeasure, therefore, a program designed to significantly increase the number of favorable genes in the human population. Muller proposed a plan called *AID*, or Artificial Insemination from Donors. He recommended the establishment of sperm banks, which would make available the sperm of highly qualified donors whose family histories showed the least possible likelihood of defects or abnormalities. To make sure of the eminence of the donors, the sperm would be frozen and made available only after 20 years or more, when the donor would no longer be alive, and posterity could judge dispassionately of his value. When a woman decided to have children, she need only choose sperm from the donor whose qualities she most admired. "How many women," Muller cried when he enunciated his plan, "would be eager and proud to bear and rear a child of Aldous Huxley or Charles Darwin!"

A number of serious questions must be answered before employing artificial insemination on the grand scale that Muller proposed. In the present state of knowledge even the best of geneticists would disagree on who should be the donors. There is no guarantee that even the most distinguished donor would be free of hidden detrimental genes. As one scientist stated, "The trouble with Muller's sperm bank is that he's always having to take people out of it." One of his rejects, for example, is Abraham Lincoln, who is now suspected of having suffered from a genetic disorder called Marfan's syndrome. Patients with Marfan's syndrome have poor musculature, long, thin extremities, and heart defects.

More important, people would be asked to surrender their pride in generating their own biological children and substitute for this pride the greater satisfaction to be gained from the knowledge that the child has the best possible set of genes. Muller confidently expected the gradual emergence of a new superior form of self-esteem and gratification in contributing to the genetic improvement of the human species. It seems more likely, however, that people would be rather uneasy at giving up their own claim to having their own biological descendants.

Other geneticists do not share Muller's pessimism that the human species is deteriorating genetically. Calculations indicate that the increase in the frequency of detrimental recessive genes due to the relaxation of selection by medical therapy would be *extremely slow*. For example, given the conditions presented in table 5.6, it would take well over 100 generations (or 3,000 years) for a recessive trait to double in frequency from 1 in 100,000 to 2 in 100,000. If the recessive defect occurred initially with a frequency of 1 in 10,000, then slightly more than 30 generations would be required to double the incidence. In essence, although relaxation of selection tends to engender an increase in the incidence of genetic defects in future generations, the increase occurs very slowly.

Table 5.6
Selection Relaxation for a Formerly Lethal Recessive Whose Fitness Is Restored to 100 Percent by Medical Therapy

Conditions: 1. Initial Frequency of Recessive Defect at Birth: 1/100,000 (1×10^{-5})
2. Mutation Rate $(A \longrightarrow a)$: 1/100,000 (1×10^{-5})

Generation	Frequency of Recessive Defect at Birth
0	1×10^{-5}
1	1.01×10^{-5}
3	1.02×10^{-5}
10	1.06×10^{-5}
30	1.20×10^{-5}
100	1.73×10^{-5}

Selected Readings

Bajema, C. J., ed. 1971. *Natural selection in human populations*. New York: John C. Wiley & Sons.

Cavalli-Sforza, L. L., and Bodmer, W. F. 1971. *The genetics of human populations*. San Francisco: W. H. Freeman and Co.

Crow, J. F. 1961. Population genetics. *American Journal of Human Genetics* 13:37–50.

Dobzhansky, T. 1950. The genetic basis of evolution. *Scientific American*, January, pp. 32–41.

———. 1961. Man and natural selection. *American Scientist* 49:285–99.

———. 1970. *Genetics of the evolutionary process*. New York: Columbia University Press.

Ehrlich, P. R., and Raven, P. H. 1969. Differentiation of populations. *Science* 165:1228–32.

Hardin, G. 1959. *Nature and man's fate*. New York: The New American Library.

Morris, L. N. 1971. *Human populations, genetic variation, and evolution*. New York: Chandler Publishing Co.

Neel, J. V., and Schull, W. J. 1968. On some trends in understanding the genetics of man. *Perspectives in Biology and Medicine* 11:565–602.

Reed, S. C. 1964. *Parenthood and heredity*. New York: John C. Wiley & Sons.

Wallace, B. 1970. *Genetic load*. Englewood Cliffs, N.J.: Prentice-Hall.

6
Selection in Action

The process of evolution is slow and continuous. We have seen that many generations of persistent selection are required to drastically reduce the frequency of an unfavorable mutant gene in a population. Likewise, it generally takes an inordinately long period of time for a new favorable mutant gene to replace its allele throughout a large population. Yet, we have encountered situations in nature in which a favorable mutation has spread through a population in a comparatively short span of years. We shall look at some outstanding examples in which we have actually observed evolution in progress.

Industrial Melanism

One of the most spectacular evolutionary changes witnessed by us has been the emergence and predominance in modern times of dark, or *melanic,* varieties of moths in the industrial areas of England and continental Europe. Slightly more than a century ago dark-colored moths were exceptional. The typical moth in the early 1800s had a light color pattern, which blended with the light coloration of tree trunks on which the moths alighted. But then the industrial revolution intervened to materially alter the character of the countryside. As soot and other industrial wastes poured over rural areas, the vegetation became increasingly coated and darkened by black smoke particles. In areas heavily contaminated with soot, the formerly abundant light-colored moths have been supplanted by the darker varieties. This dramatic change in the coloration of moths has been termed "industrial melanism." At least 70 species of moths in England have been so affected by human disturbance of the environment.

During the past two decades, several scientists, particularly E. B. Ford and H. B. D. Kettlewell at the University of Oxford, have analyzed the phenomenon of industrial melanism. Kettlewell photographed the light and dark

Figure 6.1 Dark and light forms of the peppered moth (*Biston betularia*) clinging to a soot-blackened oak tree in Birmingham, England (*left*) and to a light, lichen-coated oak tree in an unpolluted region (*right*). (Courtesy of Dr. H. B. D. Kettlewell.)

forms of the peppered moth, *Biston betularia,* against two different backgrounds (fig. 6.1). The light variety is concealed and the dark form is clearly visible when the moths rest on a light lichen-coated trunk of an oak tree in an unpolluted rural district. Against a sooty black oak trunk, the light form is conspicuous and the dark form is well camouflaged. Records of the dark form of the peppered moth date back to 1848, when its occurrence was reported at Manchester in England. At that time, the dark form comprised less than 1 percent of the population. By 1898, only 50 years later, the dark form had come to dominate the Manchester locale, having attained a remarkably high frequency of occurrence estimated at 95 percent. In fact, the incidence of the melanic type has reached 90 percent or more in most British industrial areas.

The rapid spread of the dark variety of moth is quite evident. The dark variants are protectively colored in the smoke-polluted industrial regions. They more easily escape detection by predators—insect-eating birds. Actual films taken by Kettlewell and Niko Tinbergen reveal that birds prey on the moths in a selective manner. That is, predatory birds more often capture the conspicuous light-colored moths in polluted woodlands. In a single day, the numbers of light forms in an industrial area may be reduced by as much as one-half by bird predation.

Experimental breeding tests have demonstrated that the two varieties differ principally by a single gene, with the dark variant being dominant to the light one. The dominant mutant gene was initially disadvantageous and probably was maintained at an extremely low frequency in nature by a balance between mutation and selection (see chapter 5). Then, as an indirect consequence of industrialization, the mutant gene became favored by natural selection and spread rapidly in populations in a comparatively short period of time. In unpolluted or nonindustrial areas in western England and northern Scotland, the dominant mutant gene does not confer an advantage on its bearers, and the light recessive moth remains the prevalent type.

One of the many impressive features of Kettlewell's studies lies in the unequivocal identification of the selecting agent. Selection, we may recall, has been defined as differential reproduction. The act of selection in itself does not reveal the factors or agencies that enable one genotype to leave more offspring than another. We may demonstrate the existence of selection, yet not know the cause of that selection. We might have reasonably suspected that predatory birds were directly responsible for the differential success of the melanic forms in survival and reproduction, but Kettlewell's laboriously accumulated data provided that all-important, often elusive, ingredient: *proof.*

If the environment of the peppered moth were to become altered again, then natural selection would be expected to favor the light variety again. In the 1950s, the British Parliament passed the Clean Air Act, which decreed, among other things, that factories must switch from soft high-sulfur (sooty) coal to less smoky fuels. The enforcement of this enlightened smoke-abatement law has led to a marked reduction in the amount of soot in the atmosphere. In the 1970s, the University of Manchester biologist L. M. Cook and his colleagues reported a small, but significant, increase in the frequency of the light-colored peppered moth in the Manchester area. This is further substantiation of the action and efficacy of natural selection.

Australian Rabbits

The European wild rabbit, *Oryctolagus cuniculus,* has gained infamous notoriety in Australia as the most serious economic pest ever introduced on this isolated island continent. In 1859, a small colony of 24 wild rabbits was brought from Europe to an estate in Victoria in the southeastern corner of Australia. From such modest beginnings, the rabbits multiplied enormously and by 1928 had spread over the greater part of the Australian continent. Estimates placed the number of adult rabbits at over 500 million in an area of about 1 million square miles. In overrunning the open grassy plains, the rabbits caused extensive deterioration to sheep-grazing pastures and to wheat fields.

Figure 6.2 Typical lesion that develops in a rabbit infected with myxoma virus. (Courtesy of Dr. Frank J. Fenner.)

For many years, the Australian government spent large sums of money on various measures to control the population explosion of these prolific rabbits. Trapping, rabbit-proof fencing, poisoning of water holes, and fumigation all proved to be largely ineffectual. Then, beginning in 1950, outstanding success in reducing the rabbit population was achieved by inoculating rabbits with a virus that causes the fatal disease *myxomatosis*. The deadly myxoma virus was implanted into the tissues of rabbits in the southern area of Australia (fig. 6.2). In a remarkably short period of time, the virus had made its way, aided by insect carriers (mosquitoes), into most of the rabbit-infested areas of the continent. By 1953, more than 95 percent of the rabbit population in Australia had been annihilated.

However, after their drastic decline in the early 1950s, the rabbit populations began to build up again. Evolutionary changes have occurred in both the pathogen (virus) and the host (rabbit). Mutations conferring resistance to the myxoma virus have selectively accumulated in the rabbit populations. At the same time, the viruses themselves have undergone genetic changes; less virulent strains of the virus have evolved.

Selection would favor the emergence of less virulent strains of a pathogen leading to less severe expressions of the disease. The advantage to the pathogen is comprehensible, since a pathogen which would cause the instantaneous death of the host also would cause its own prompt demise from lack of an immediate host. Experiments by the Australian scientist Frank J. Fenner in the 1950s showed that new strains of the myxoma virus had arisen. Highly susceptible, standard laboratory rabbits were infected with different preparations that had been isolated in successive epidemics in Australia. The early, or original, virus strains caused greater mortality in the laboratory rabbits than strains taken from the later epidemics. With each successive epidemic, a less virulent strain had evolved. In nature, the original, highly virulent strain of virus had killed the rabbits within a few days, which limited the virus's chances of being transmitted to a new host through the mosquito vector. However, when a virus of lower virulence appeared, the duration of the disease in the rabbit was extended, with the consequence that the less virulent mutant strain improved its chances of being transmitted to unaffected rabbits through the bite of a mosquito. Accordingly, attenuated, or weakened, strains of the virus progressively displaced in nature the original highly virulent strain.

Experiments by Fenner and his collaborators also demonstrated that the rabbits had become genetically resistant to the virus. Uniform doses of a standard virus preparation were injected into laboratory offspring of rabbits that had been trapped in successive epidemics. Significantly greater numbers of experimental offspring of parents from the first drastic epidemic succumbed to the inoculated virus compared with offspring of parents from the later, relatively mild epidemics. Thus, successive epidemics fostered the accumulation of genes protecting the rabbits against death from the myxoma virus. Stated another way, genetically susceptible rabbits had died, leaving the more genetically resistant rabbits to contribute more offspring each generation. In essence, natural selection has favored a mutual accommodation between the virus and the Australian rabbit.

Selection for Resistance

Penicillin, sulfonamides (sulfa drugs), streptomycin, and other modern antibiotic agents made front-page headlines when first introduced. These wonder drugs were exceptionally effective against certain disease-producing bacteria, and contributed immeasurably to the saving of human lives in World War II. However, the effectiveness of these drugs has been reduced by the emergence of resistant strains of bacteria. Medical authorities regard the rise of resistant bacteria as the most serious development in the field of infectious diseases over the past decade. Bacteria now pass on their resistance to antibiotics faster than people spread the infectious bacteria.

Mutations have occurred in bacterial populations that enable the mutant bacterial cells to survive in the presence of the drug. Here again we notice that mutations furnish the source of evolutionary changes, and that the fate of the mutant gene is governed by selection. In a normal environment, mutations that confer resistance to a drug are rare or undetected. In an environment changed by the addition of a drug, the drug-resistant mutants are favored and supplant the previously normal bacterial strains.

It might be thought that the mutations conferring resistance are actually caused or induced by the drug. But this is not the case. Drug-resistant mutations arise in bacterial cells irrespective of the presence or absence of the drug. An experiment devised by the Stanford University geneticist Joshua

Lederberg provides evidence that the drug acts as a selecting agent, permitting preexisting mutations to express themselves. As seen in figure 6.3, colonies of bacteria were grown on a streptomycin-free agar medium in a petri plate. When the agar surface of this plate was pressed gently on a piece of sterile velvet, some cells from each bacterial colony clung to the fine fibers of the velvet. The imprinted velvet could now be used to transfer the bacterial colonies onto a second agar plate. In fact, more than one replica of the original bacterial growth can be made by pressing several agar plates on the same area of velvet. This ingenious technique has been appropriately called *replica plating*.

In preparing the replicas, Lederberg used agar plates containing streptomycin. Of course, on these agar plates, only bacterial colonies resistant to streptomycin grew. In the case depicted in figure 6.3, one colony was resistant. Significantly, this one resistant colony was found in the same exact position in all replica plates. If mutations arose in response to exposure to a drug, it is hardly to be expected that mutant bacterial colonies would arise in precisely the same site on each occasion. In other words, a haphazard or random distribution of resistant bacterial colonies, without restraint or attention to location in the agar plate, would be expected if the mutations did not already exist in the original bacterial colonies.

Now, we can return to the original plate, as Lederberg did, and test samples of the original bacterial colonies in a test tube for sensitivity or resistance to streptomycin (bottom part of fig. 6.3). It is noteworthy that the bacterial colonies on the original plate had not been previously in contact with the drug. When these original colonies were isolated and tested for resistance to streptomycin, only one colony proved to be resistant. This one colony occupied a position on the original plate identical with the site of the resistant colony on the replica plates. The experiment demonstrated conclusively that the mutation had not been induced by streptomycin but had already been present before exposure to the drug.

Other genetic changes perpetuated by selection have been detected in recent years. The insecticide with the awesome name of dichloro-diphenyl-trichloroethane, or DDT, was initially highly poisonous to malaria-carrying mosquitoes and typhus-carrying lice. However, in the three decades since the introduction of this insect poison, numerous mutant DDT-resistant strains of mosquitoes and lice have appeared in widely different parts of the world. Similarly, hydrogen cyanide has for many years been an efficient fumigant in the destruction of insect pests, particularly the scale insect, in the citrus-producing areas of southern California. Mutant strains of scale insects that can tolerate the usual lethal doses of hydrogen cyanide have recently arisen in several of the fruit orchards of California. Our struggle against diseases and insect pests is unending as long as the mutation process continually produces variants that may have survival value in an altered environment.

Figure 6.3 **Experiment by Joshua Lederberg** revealing that drug-resistant mutations in bacterial cells had not been induced by the drug but had already been present prior to exposure of the bacteria to the drug.

SPONTANEOUS ORIGIN OF MUTATION

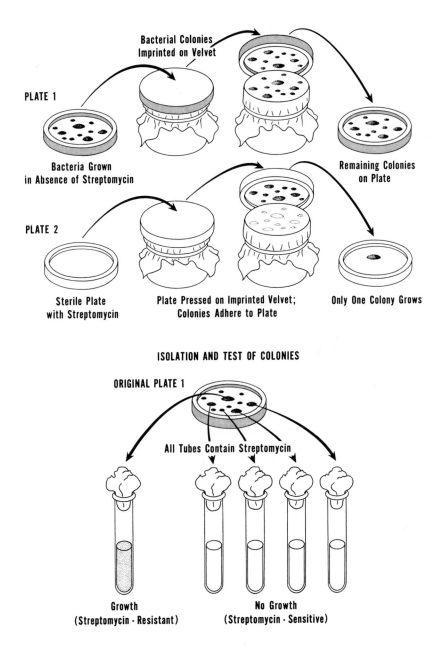

Bacterial Colonies
Imprinted on Velvet

PLATE 1

Bacteria Grown
in Absence of Streptomycin

Remaining Colonies
on Plate

PLATE 2

Sterile Plate
with Streptomycin

Plate Pressed on Imprinted Velvet;
Colonies Adhere to Plate

Only One Colony Grows

ISOLATION AND TEST OF COLONIES

ORIGINAL PLATE 1

All Tubes Contain Streptomycin

Growth
(Streptomycin - Resistant)

No Growth
(Streptomycin - Sensitive)

Parenthetically, it should be noted that DDT has fallen into increasing disfavor as a threat to wildlife. DDT is an extremely stable compound, retaining its chemical potency for many years rather than breaking down harmlessly within a few days after application. Long-lasting DDT residues have spread within two decades throughout the world by wind and water in much the same way as radioactive fallout. DDT has been found in areas of the globe where the insecticide has never been used. Penguins and crab-eating seals in the Antarctic, far removed from any actual spraying, have accumulated traces of the compound. It has been estimated that there are approximately one billion pounds of DDT currently circulating in the biosphere.

The DDT levels in the gonads and eggs of several species of birds have seriously impaired the reproductive potential of these organisms. Since DDT interferes with the mobilization of calcium in the oviduct, the eggs are laid with exceedingly thin shells, subject to hairline cracks, or with no shell at all. Birds of prey, such as the bald eagle and osprey, are now on the brink of extinction. Whereas ospreys formerly averaged 2.5 young a year, today they are barely able to produce 1 young a year. As William Mudock of the University of California (Santa Barbara) stated, "no amount of technological ingenuity can reconstruct a species of osprey once it is extinct." Even with the curbed use of DDT in several countries (Sweden, Canada, and the United States), it would take several decades for the environment to purge itself of this extremely long-lasting pollutant.

Infectious Diseases in Humans

The British geneticist J. B. S. Haldane has suggested that infectious diseases have been one of the most potent agents of natural selection in human populations. It is difficult to prove unequivocally that genetic changes have occurred in the human resistance to pathogens during widespread epidemics in the past. Nevertheless, it is a reasonable supposition that major epidemics eliminated a large percentage of genetically susceptible individuals, and permitted the selective survival and multiplication of individuals endowed with greater genetic resistance.

Up until 10,000 years ago, humans subsisted by food-gathering and hunting alone. Widely dispersed groups of nomadic peoples lived at low population densities. Contact between groups was limited, and contagious diseases occurred infrequently. With the development of agriculture and the establishment of village settlements, human populations became dense and concentrated in relatively small regions. This set the stage for infectious community diseases, spread by contact, such as tuberculosis and bubonic plague, both caused by bacteria. Viral diseases, such as measles, mumps, chickenpox, and smallpox, posed major threats to the health of preindustrial humans. With the advent of the industrial revolution, the densely populated urban areas witnessed repeated outbreaks of infectious diseases. In industrial times, many diseases—in particular, tuberculosis—flared up to ravage entire populations.

An infectious disease tends to reach epidemic proportions in populations that have had no previous exposure to the disease. When populations are first exposed to a contagious disease, the mortality is high and the infection is acute. On evolutionary principles, repeated exposure over many generations should lead to a genetic reconstruction of the population, with a raised frequency of genetically resistant individuals. Thus, in subsequent episodes of the disease, the expectations are lowered mortality and less severe manifestations of the disease.

Strong presumptive evidence that heritable resistance can arise through selection is furnished by the change, with time, in the incidence and character of tuberculosis among the Plains Indians of the Qu'Appelle Valley Reservation in Saskatchewan, Canada. In 1881, the Plains Indians, shortly after arriving at the reservation, became exposed to the tuberculosis germ for the first time. By 1886, almost everyone in the reservation had contracted the infection. The annual death rate from tuberculosis in 1886 reached the incredible figure of 900 per 10,000, or approximately 10 percent of the total population. The disease was not only widespread but also highly virulent, causing damage to bone, meninges, kidney, and other parts of the body. Such multiple disorders reflect the inability of the host to localize the infection. In 1921, however, the disease showed a pronounced tendency to localize in the lungs and to pursue the chronic course characteristic of pulmonary tuberculosis. Mortality dropped to 7 percent among Indian school children in 1921, and in 1950, mortality declined to 0.2 percent. Accordingly, in but a few generations, less than one percent of the Indian children died of tuberculosis. It should be stressed that this dramatic decline occurred in absence of specific chemotherapy and modern public health measures.

Other communities of the world have shown the same pattern of evolutionary change. Europeans typically exhibit a mild form of tuberculosis. René Dubos, a microbiologist at the Rockefeller Institute, carefully analyzed the available data, and concluded that Europeans today are the beneficiaries of an exacting selective process brought about by repeated outbreaks of tuberculosis a few generations ago. During the eighteenth and nineteenth centuries, tuberculosis spread with the growth of cities, and created havoc throughout the industrialized world. It became known as the "Great White Plague." Annual mortality rates during the industrial era were as high as 500 per 100,000.

Tuberculosis is a highly contagious disease. Minute droplets of sputum, discharged by the cough or sneeze of an affected person, contain literally thousands of tubercle bacteria, which can survive in the air for several hours. Even thoroughly dried sputum, when it is inhaled, can be infective. Tuberculosis flared in crowded London in 1750, and then swept uncontrollably through the large cities of continental Europe. Immigrants from Europe brought the devastating disease to the United States in the early 1800s. The death toll was highest in cities with large immigrant populations—New York,

Philadelphia, and Boston. As tuberculosis advanced across the nation, it became the chief cause of death in the United States in the nineteenth century. Tuberculosis then declined steadily in the twentieth century in all industrialized countries. From a high value of 500 deaths per 100,000 in the nineteenth century, the mortality rate dropped in Europe and the United States to approximately 190 per 100,000 in 1900. By 1945, the rate fell to about 40 per 100,000.

The decline in mortality cannot be attributable, in any large measure, to general advances in living conditions, or to therapy. The decline was in full force before effective medical steps were taken, and indeed, even before the discovery of the tubercle bacteria. The causative agent was discovered only in 1882 by Robert Koch in Germany. The available information suggests strongly that the reduction in mortality has primarily been the result of an evolutionary attenuation of the virulence of the pathogen coupled with a heightened, genetically conditioned, resistance level of the host. In other words, the severity of infection has most likely been diminished by genetic selection and has been essentially independent of medical or public health intervention.

It is only since the 1940s that the benefits of chemotherapy have become detectable in the Western world. Today, cases of tuberculosis occur at an annual rate of 20 per 100,000, and most cases are permanently arrested by treatment with such drugs as isonicotinic acid hydrazide (INH) and para-amino salicylic acid (PAS). The effectiveness of modern drugs seems to indicate that populations of parts of the world in which tuberculosis still prevails can avoid the remorseless selective process in which death eliminates the susceptible genotypes. Yet, some 3 million persons each year succumb to tuberculosis in Latin America, Africa, and Asia. Although effective drugs to combat tuberculosis are available for worldwide distribution, most of the poorly developed nations have inadequate medical and health services, and large proportions of their populations have only remote contact with any medical care.

As the prevalence of tuberculosis has decreased with the use of chemotherapeutic agents, so has the threat to genetically susceptible individuals. Persons who earlier would have succumbed to tuberculosis are now surviving. In other words, the reproductive fitness of genetically susceptible individuals has increased. Reduction in selection pressure for disease resistance has probably been accompanied by an increase in the frequency of genetic factors for susceptibility to tuberculosis. Since other animal species continue to harbor the tuberculosis germ, human populations would be especially vulnerable to outbreaks of tuberculosis in the event of lapses in public health measures.

General Effects of Selection

In the foregoing discussion, we have seen that the selection process favors individuals that are best adapted to new situations or to new ecological opportunities. Such selection is said to be *directional* since the norm for the

Figure 6.4 Schematic representation of three types selection and their effects. Each curve represents the normal distribution of the character in a population. From top to bottom, the lower curves show the expected distribution after the impact of selection. A. *Stabilizing selection.* The intermediate values for a given trait are favored (preserved) and the shape of the curve narrows through the elimination of the carriers of the extreme values. B. *Directional selection.* The adaptive norm changes as less adapted genotypes are replaced by better adapted genotypes. C. *Disruptive selection.* Two adaptive norms are generated when the population exists in a heterogeneous environment.

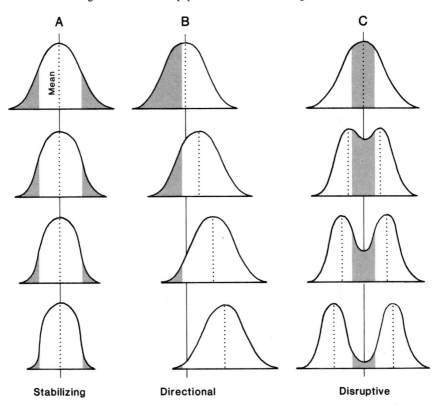

population is shifted with time in one direction (fig. 6.4*B*). The selection for low virulence that occurred over time among the myxoma viruses in the host rabbits exemplifies directional selection. The raw materials for selection were a series of gene mutations, each of which acted to reduce the virus's virulence. The curve of distribution shifted to the right (fig. 6.4*B*) as the less harmful strains of viruses displaced the virulent strains.

The selection process may take other forms. *Stabilizing* selection acts to reduce the array of gene complexes that can be expressed in a population. As seen in figure 6.4*A*, the shape of the curve tends to narrow through the continual elimination of the less adapted individuals at the extremes of the distribution curve. Stated differently, the intermediate values for a given trait

Figure 6.5 The distribution of birth weights of human newborns and the mortality of the various birth-weight classes. The histogram *(left)* shows the proportions of the population falling into the various birth-weight classes. The mean birth weight is 7.1 pounds. The curve of mortality *(right)* in relation to birth weight reveals that the lowest mortality is associated with the optimum birth weight (7.3 pounds). (Based on data by Karn and Penrose, 1951.)

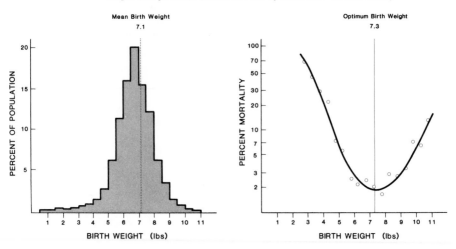

are favored over the extreme values. The birth weight of newborns provides an instructive example of a human character that has been subjected to stabilizing selection (fig. 6.5). The optimum birth weight is 7.3 pounds; newborn infants less than 5.5 pounds and greater than 10 pounds have the highest probability of mortality. Given the strong stabilizing influence of weeding out the extremes, the optimum birth weight is associated with the lowest mortality. The curve for mortality is virtually the complement of the curve for survival.

Disruptive selection is the most unusual of the three types of selection (fig. 6.4C). This form of selection occurs when the extreme values have the highest fitness and the intermediate values are relatively disadvantageous in terms of reproductive effectiveness. It is, essentially, selection for diversification with respect to a character. The shell patterns of limpets (marine molluscs) form a continuum ranging from pure white to dark tan. Limpets typically dwell in one of two distinct habitats, attaching either to white goose-neck barnacles, or to tan-colored rocks. As might be expected, the light-colored limpets seek the protection of the white barnacles, whereas tan limpets live by choice almost exclusively on the dark rocks. Limpets of intermediate shell patterns are conspicuous and are intensely selected against by predatory shore birds. If this disruptive type of selection (favoring the extremes) were to be accompanied by the sexual isolation of the two types of limpets, then two new species could arise. The mode of origin of species is treated in detail in a subsequent chapter.

Selected Readings

Bishop, J. A., and Cook, L. M. 1975. Moths, melanism and clean air. *Scientific American,* January, pp. 90–99.

Dubos, R. 1968. *So human an animal.* New York: Charles Scribner's Sons.

Dubos, R., and Dubos, J. 1952. *The white plague: tuberculosis, man and society.* Boston: Little, Brown and Co.

Fenner, F. 1959. Myxomatosis in Australian wild rabbits—evolutionary changes in an infectious disease. *The Harvey Lectures 1957–58.* New York: Academic Press.

Ford, E. B. 1960. *Mendelism and evolution.* London: Methuen & Co.

Harrison, G.; Weiner, J.; Tanner, J.; and Barnicot, N. 1964. *Human biology: an introduction to human evolution.* New York: Oxford University Press.

Huxley, J. 1953. *Evolution in action.* New York: New American Library of World Literature.

Kettlewell, H. B. D. 1959. Darwin's missing evidence. *Scientific American,* March, pp. 48–53.

Motulsky, A. G. 1960. Metabolic polymorphisms and the role of infectious diseases in human evolution. *Human Biology* 32:28–62.

Neel, J. V. 1958. The study of natural selection in primitive and civilized human populations. *Human Biology* 30:43–72.

Ryan, F. J. 1953. Evolution observed. *Scientific American,* October, pp. 78–83.

Tax, S., ed. 1960. *Evolution after Darwin* (3 vols.). Chicago: University of Chicago Press.

7
Balanced Polymorphism

The concepts presented in the preceding chapter have led us to believe that selection operates at all times to reduce the frequency of an abnormal gene to a low equilibrium level. This view is not entirely accurate. We are aware of genes with deleterious effects that occur at fairly high frequencies in natural populations. A striking example is the high incidence in certain human populations of a mutant gene that causes a curious and usually fatal form of blood cell destruction, known as *sickle-cell anemia*. It might be presumed that this detrimental gene is maintained at a high frequency by an exceptionally high mutation rate. There is, however, no evidence to indicate that the sickle-cell gene is usually mutable. We now know that the maintenance of deleterious genes at unexpectedly high frequencies involves a unique, but not uncommon, selective mechanism, which results in a type of population structure known as *balanced polymorphism*.

Sickle-cell Anemia

This disease was discovered by the American physician James B. Herrick, who in 1904 made an office examination of an anemic West Indian black student residing in Chicago. The patient's blood examined under the microscope showed the presence of numerous crescent-shaped erythrocytes. The peculiarly twisted appearance of the red blood cell is shown in figure 7.1. The black patient was kept under observation for six years, during which time he displayed many of the distressing symptoms we now recognize as typical of the disease. The bizarre-shaped red cells in the form of a sickle blade are fragile and clog small blood vessels. The obstruction of circulation leads in turn to the necrosis (death) of various tissues. The damage to organs (heart, lungs, and kidneys) is accompanied by a variety of complications, including pneumonia, rheumatism, leg ulcers, and heart failure.

Figure 7.1 **Peculiarly twisted shape** of red blood cells of an individual suffering from sickle-cell anemia (*top*), contrasted with the spherical appearance of normal red blood cells (*bottom*). (Wide World Photos.)

The disease is variable in severity. A few persons are only mildly affected, while many are seriously disabled and fail to survive to reproductive age. Physicians can provide care for the patient, but there is no cure. In other words, there is little that can be done for sickle-cell patients beyond treatment of the symptoms. Intense episodes of localized pain, or "crises," are the hallmark of sickle-cell anemia. Although the crises may occur in any part of the body, they usually arise in the joints, back, chest, and abdomen. The sharp crises may last only a few hours, or endure for two or three weeks. Moderate clinical success in reversing the sickling of blood cells has been achieved by the use of several chemical agents—urea, cyanate, zinc, and recently, compounds called amino acid benzyl esters. Despite chemical therapy, the crises still linger and damage to organs, although abated, persists.

Sickle-cell anemia occurs predominantly in blacks. It is about six times more common than the next most common long-term illness of black children (diabetes). The incidence at birth of the disabling sickle-cell anemia in the United States is estimated at 2 per 1,000 infants. Not until after World War II was the hereditary basis of sickle-cell anemia uncovered. The irregularity was shown in the late 1940s by James V. Neel of the University of Michigan to be inherited as a simple Mendelian character. The sickle-cell anemic patient inherits two defective genes, one from each parent (fig. 7.2). Individuals with one normal and one defective (sickling) gene are generally healthy but are carriers—they have *sickle-cell trait*. If two heterozygous carriers marry, the chances are one in four that a child will have sickle-cell anemia, and one in two that a child will be a carrier (fig. 7.2).

Since the homozygous state of the detrimental gene is required for the overt expression of the disease, sickle-cell anemia may be considered to be recessively determined. However, although heterozygous individuals are, on the whole, normal, even the red cells of the heterozygotes can undergo sickling under certain circumstances, producing clinical manifestations. As an instance, heterzygous carriers have been known to experience acute abdominal pains at high altitudes in unpressurized airplanes. The lowering of the oxygen tension, with an increased amount of oxygen released from chemical combination in the erythrocytes, is sufficient to induce sickling. The abdominal pains can be traced to the packing of sickled erythrocytes in the small capillaries of the spleen.

According to some authors, since the detrimental gene can express itself in the heterozygous state (by producing a positive sickling test under certain circumstances), the gene should be considered dominant. This reveals that "dominance" and "recessiveness" are somewhat arbitrary concepts that depend on one's point of view. Indeed, from a molecular standpoint, the relation between the normal and defective gene in this instance may best be described as *codominant,* since, as we shall see below, the heterozygote produces both normal and abnormal hemoglobin. There is no blending of inheritance at the molecular level.

Sickle-cell Hemoglobin

In 1949, the distinguished chemist and Nobel laureate Linus Pauling and his co-workers made the important discovery that the defective sickling gene alters the configuration of the hemoglobin molecule. Pauling used the then relatively new technique of electrophoresis, which characterizes proteins according to the manner in which they move in an electric field. The hemoglobin molecule travels toward the positive pole. Hemoglobin from a sickled cell differs in speed of migration from normal hemoglobin; it moves more slowly than the normal molecule (fig. 7.3). The defective gene thus functions differently from the normal gene, and, in fact, acts independently of its normal partner allele. As a result, the heterozygote does not produce an intermediate product, but instead produces both kinds of hemoglobin in nearly equal quantities—the normal type (designated *hemoglobin A,* or *Hb A*) and the sickle-cell anemic variety *(Hb S).* The dual electrophoretic pattern of hemoglobin from a het-

Figure 7.2 **Types of children** that can result from a marriage of two heterozygous carriers of sickle-cell anemia. Individuals homozygous for the variant
gene suffer from sickle-cell anemia; the benign heterozygous state is referred
to as the sickle-cell trait.

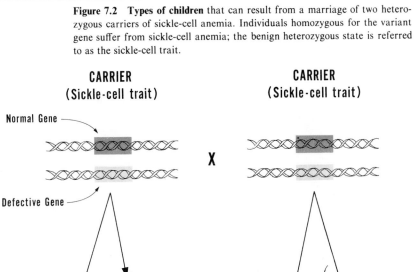

Figure 7.3   Electrophoretic patterns of hemoglobins. The hemoglobin of the heterozygous person with sickle-cell trait is not intermediate in character, but is composed instead of approximately equal proportions of the normal hemoglobin and the sickle-cell anemia variety. (Based on studies by Linus Pauling, in Pauling, Itano, Singer, and Wells, "Sickle-Cell Anemia: A Molecular Disease," *Science*, vol. 110, 25 November 1949, pp. 543–48.)

HEMOGLOBIN ELECTROPHORETIC PATTERN

erozygous individual can actually be duplicated experimentally by mechanically mixing the hemoglobins taken from blood cells of a normal person and a sickle-cell anemic patient. The mixed solution separates in an electric field into the same two hemoglobin components as those characteristic of a heterozygous person with sickle-cell trait.

The sickle-cell mutant gene modifies the molecular structure of hemoglobin, and sickle-cell anemia has been appropriately described as a *molecular disease*. It remained for the chemist Vernon Ingram, than at Cambridge University in England, to discover how the hemoglobin molecule is altered by the aberrant gene. In 1956, Ingram succeeded in breaking down hemoglobin, a large protein molecule, into several peptide fragments containing short sequences of identifiable amino acids. Normal hemoglobin and sickle-cell hemoglobin yielded the same array of peptide fragments, with a single exception. In one of the peptide fragments of sickle-cell hemoglobin, the amino acid *glutamic acid* had been replaced at one point in the chain by *valine* (fig. 7.4). The sole difference in chemical composition between normal and sickle-cell hemoglobin is the incredibly minute substitution of a single amino acid unit among several hundred. The fatal effect of sickle-cell anemia is thus traceable to an exceedingly slight alteration in the structure of a protein molecule. The mutation itself represents a highly localized change in one of the nucleotide pairs in the chromosomal DNA molecule (see fig. 3.5).

Gene Action

It has been said that hemoglobin served as the Rosetta stone in revealing the hereditary language. It was the hemoglobin molecule that provided the first direct proof of the relationship of the gene to protein synthesis. Proteins are not synthesized directly on the DNA template. The genetic information (genetic code) of DNA is first transcribed to a secondary form of nucleic acid,

Figure 7.4   Amino acid sequences in a small section of the normal hemoglobin molecule and of the sickle-cell hemoglobin. The substitution of a single amino acid, glutamic acid by valine, is responsible for the abnormal sickling of human red blood cells. (Based on studies by Vernon Ingram.)

Normal Hemoglobin

valine   histidine   leucine   threonine   proline   glutamic acid   glutamic acid   lysine

Sickle-Cell Hemoglobin

valine

ribonucleic acid, or RNA (fig. 7.5). The RNA molecule resembles the DNA molecule in structure except in three important respects. The pyrimidine base, *uracil,* is found in RNA, which replaces the thymine that is characteristic of DNA. Secondly, the sugar in RNA is *ribose,* which contains one more oxygen atom than does the deoxyribose sugar. Thirdly, RNA has only a *single* strand instead of two.

As seen in figure 7.5, one of the two strands of DNA forms a complementary strand of RNA. The same rules of pairing hold as in replicating a copy of DNA, except that adenine attracts uracil instead of thymine. This RNA strand, which is responsible for carrying DNA's instructions out into the cytoplasm, is appropriately termed *messenger RNA, or mRNA.* The amino acids of a polypeptide chain are specified by the messenger RNA. In a large sense then, the genetic code applies to RNA rather than to DNA itself. The transfer of information from DNA to messenger RNA has been called *transcription.* The transfer of information from messenger RNA to proteins is termed *translation* (fig. 7.5). The manner in which messenger RNA directs the assembly into amino acids into polypeptide chains is beyond the scope of this book.

Scientists have broken the genetic code into its main elements. Each 3-letter unit of the messenger RNA is called a *codon.* Each of the twenty main amino acids, ranging from alanine to valine, has been shown to be specified by at least one codon. In fact, most amino acids have more than one codon. The amino acid serine, for example, has six codons, glycine has four, and lysine has two. Only methionine and tryptophan have one each. Because most amino acids are specified by more than one codon, the code is claimed to be degenerate. However, this kind of degeneracy is welcome since it ensures that the code works. In normal hemoglobin, glutamic acid is specified by the DNA triplet CTT, which in turn is coded by the messenger RNA codon, GAA. A single base substitution, resulting in the triplet CAT (rather than CTT) and the codon GUA (rather than GAA), accounts for the presence of valine instead of glutamic acid at a specific site in sickle cell hemoglobin.

Figure 7.5 The language of life. The DNA molecule forms a single strand of messenger RNA that carries DNA's instructions out into the cell. The four bases of DNA (A, T, C, and G) are responsible for the 3-letter code words, or codons, of messenger RNA. Each of the twenty amino acids that make up the variety of body proteins is specified by at least one codon.

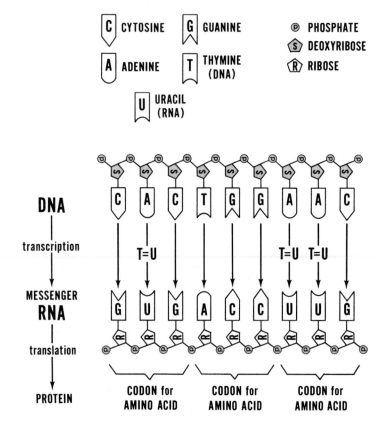

One of the most intriguing findings is that the genetic code is essentially universal. The same codon calls forth the same amino acid in organisms as widely separated phylogenetically as a bacterium *(E. coli)*, a flowering plant (wheat), an amphibian (the South African clawed toad), and mammals (guinea pig, mouse, and human). The essential universality of the genetic code supports the view that the code had its origin at least by the time bacteria evolved, 3 billion years ago. The genetic code may be said to be the oldest of languages.

Theory of Balanced Polymorphism

Since persons with sickle-cell anemia ordinarily do not survive to reproductive age, it might be expected that the detrimental gene would pass rapidly from existence. Each failure of homozygous anemic individual to transmit his genes would result each time in the loss of two aberrant genes from the population. And yet, the sickle-cell gene reaches remarkably high frequencies in the tropical zone of Africa. In several African populations, 20 percent or more of the individuals have the sickle-cell trait, and frequencies as high as 40 percent have been reported for some African tribes. The sickle-cell trait is not confined to the African continent; it has been found in Sicily and Greece, and in parts of the Near East. What can account for the high incidence of the sickle-cell gene, particularly in light of its detrimental action?

As an aid in answering the foregoing question, we will consider the sickling gene to be recessive. Accordingly, we can symbolize the normal gene as *A* and its allele for sickling as *a*. Now, the explanation for the high level of the deleterious sickle-cell gene is to be found in the possibility that the heterozygote *(Aa)* is superior in fitness to *both* homozygotes (*AA* and *aa*). In other words, selection favors the heterozygote and both types of homozygotes are relatively at a disadvantage. Let us examine the theory behind this form of selection.

Figure 7.6 illustrates the theory. The classical case of selection discussed in preceding chapters is portrayed in the first part of the figure. In this case, when *AA* and *Aa* individuals are equal in reproductive fitness, and the *aa* genotype is completely selected against, the recessive gene will be eliminated. Barring recurring mutations, only *A* genes will ultimately be present in the population.

Figure 7.6 Equilibrium frequencies of two alleles (*A* and *a*) under different conditions of selection (ignoring mutation). In contrast to case I (complete selection and total elimination of *a*), the recessive gene can be retained at appreciable frequencies in a population when the heterozygote *(Aa)* is superior in reproductive fitness to *both* homozygotes (cases II, III, and IV, showing different relative fitnesses of the two homozygotes, *AA, and aa*).

Let us assume in another situation (case II in fig. 7.6) that both homozygotes (*AA* as well as *aa*) are incapable of leaving surviving progeny. The only effective members in the population are the heterozygotes *(Aa)*. Obviously, the frequency of each gene, *A* and *a,* will remain at a constant level of 50 percent. The inviability of both homozygous types probably never exists in nature, but the scheme does reveal how a stable relation of two alleles at high frequencies is possible. If, as in the case III, the *AA* genotype leaves only half as many progeny as the heterozygote, and the recessive homozygote is once again inviable, then it is apparent that more *A* alleles are transmitted each generation than *a* alleles. Eventually, however, the *A* gene will reach an equilibrium point at 0.67. Here, it may be noted that, although the recessive homozygote is lethal, the frequency of the recessive allele *(a)* is maintained at 0.33. In the last illustrative example (case IV), the recessive homozygote is not as disadvantageous as the dominant homozygote, but both are less reproductively fit than the heterozygote. Here also, both genes remain at relatively high frequencies in the population. Indeed, the recessive allele *(a)* will constitute 67 percent of the gene pool.

We have thus illustrated in simplified form the selective forces that serve to maintain two alleles at appreciable frequencies in a population. This phenomenon is known as *balanced polymorphism*. The loss of a deleterious recessive gene through deaths of the homozygotes is balanced by the gain resulting from the larger numbers of offspring produced by the favored heterozygotes.

Experimental Verification of Balanced Polymorphism

A simple experiment, using the fruit fly, *Drosophila melanogaster,* can be performed to demonstrate that an obviously lethal gene can be maintained at a stable, relatively high frequency in a population. The geneticist P. M. Sheppard of the University of Liverpool in England introduced into a breeding cage a population of flies of which 86 percent were normal and 14 percent carried the mutant gene *stubble,* which affects the bristles of the fly. The stubble gene is lethal when homozygous; hence, all stubble individuals are heterozygous. Ordinarily the heterozygous fly does not have any reproductive advantage over the normal homozygote. Sheppard, however, created a situation whereby the heterozygote would be favored by removing 60 percent of the normal flies from the population each generation. Consequently, the heterozygote was rendered superior in fitness to either homozygote by virtue of the natural lethality of the stubble homozygote and the enforced reproductive incapacity of many of the normal homozygotes. The results of the experiment are recorded in table 7.1.

Table 7.1
Experimental Balanced Polymorphism

Generation	Percentage of Stubble Flies	Frequency of Stubble Gene
1	14.3	0.0715
2	33.7	0.1685
3	57.6	0.2880
4	63.2	0.3160
5	69.1	0.3455
6	73.5	0.3675
7	72.9	0.3645
8	73.4	0.3670
9	72.9	0.3645

From P. M. Sheppard, *Natural Selection and Heredity*. New York: Philosophical Library, 1959.

It should be noted that the frequency of the stubble gene increased in the early generations, but then became stabilized at a level of about 0.365. The equilibrium level is reached when as many stubble genes are lost from the population, through death of the stubble homozygotes, as are gained as a result of the reproductive advantage of the heterozygote. Although the stubble gene is lethal, the population, under the constant conditions of the experiment, remained at a stable state with a high number of heterozygotes.

The frequency of the stubble gene will fall rapidly when the usual reproductive potential of the normal homozygote is restored. When the normal homozygotes and heterozygotes are equal in reproductive fitness (refer to case I, fig. 7.6), the normal gene will supplant the abnormal stubble gene in the population. With these considerations in mind, we shall return to our discussion of sickle-cell anemia.

Superiority of the Heterozygotes

The high frequency of the sickle-cell gene in certain African populations can be explained by assuming that the heterozygotes (individuals with sickle-cell trait) have a selective advantage over the normal heterozygotes. What might be the nature of the advantage? Field work undertaken in Africa in 1949 by the British geneticist Anthony Allison revealed that the incidence of the sickle-cell trait is high in regions where malignant subtertian malaria caused by the parasite *Plasmodium falciparum* is hyperendemic—that is, transmission of the infection occurs throughout most of the year. Thus, the population is almost constantly reinfected with malaria, and susceptible individuals suffer from severe and prolonged attacks of malaria. Under such circumstances, relative immunity to falciparum malaria would be most beneficial.

Allison examined blood from African children and found that carriers of the sickle-cell trait were relatively resistant to infection with *Plasmodium falciparum*. The heterozygous carriers were infected less often with the parasite

than the homozygous dominant nonsicklers. Moreover, among those heterozygotes that were infected, the incidence of severe, or fatal, attacks of malaria
was strikingly low. The evidence is strong that the sickle-cell gene affords some
degree of protection for young children against malarial infection. Hence, in
areas where malaria is common, children possessing the sickle-cell trait will
tend to survive more often than those without the trait, and are more likely
to pass on their genes to the next generation. The heterozygotes *(Aa)* are thus
superior in fitness to both homozygotes, which are likely to succumb either
from anemia on the one hand *(aa)* or malaria on the other *(AA)*.

Frank B. Livingstone of the University of Michigan has impressively revealed the real relationship among the spread of agriculture, malaria, and
sickle-cell anemia. The spread of the sickling gene was greatly enhanced by
the development of agriculture. The clearing of the forest in the preparation
of ground for cultivation provided new breeding areas for the mosquito
Anopheles gambiae, the vector, or carrier, of the malaria plasmodium. The
spread of malaria has been responsible for the spread of the selective advantage of the sickle-cell gene, which in the heterozygous state imparts resistance
to malaria. In essence, the selective advantage of the heterozygote tends to
increase in direct proportion to the amount of malaria present in a given area.
Hunting populations in Africa show a very low incidence of malaria and an
equally low frequency of the sickling gene. The Pygmies of the Ituri Forest
constitute a good example.

Communities in Africa with the greatest reliance on agriculture (rather
than on hunting or animal husbandry) tend to have the highest frequencies of
the sickle-cell trait. A high incidence of the sickle-cell trait in an intensely
malarious environment has the consequence of reducing the number of individuals capable of being infected by the malarial parasite and, accordingly, of
lowering the mortality from such infections. More human energy, or greater
manpower, is thus made available for raising and harvesting crops. Ironically,
then, the sickle-cell trait carries with it the beneficial effect of enabling tribes
to develop and maintain an agricultural culture rather than adhere to a hunting
or pastoral existence. This is a curious, but striking, instance of the interplay
of biological change and socioeconomic adaptation.

In a similar finding, evidence exists that persons whose red blood cells are
deficient in the enzyme G6PD (glucose-6-phosphate-dehydrogenase) are less
likely to be affected by malaria. Data also show a strong correlation between
the incidence of malaria and the frequency of thalassemia (a type of anemia)
in the Italian peninsula and in Sardinia. Malaria apparently has had a profound influence on human events.

Disruption of Balanced Polymorphism

We should expect the frequency of the sickle-cell gene to be low in malaria-
free areas, where the selective advantage of the heterozygote would be removed. Accordingly, we find that the lowest frequencies of the sickle-cell gene
occur consistently in regions relatively free of malaria. The frequency of the

sickle-cell gene has fallen to relatively low levels in the black populations of the United States. The frequency of the sickle-cell trait among American blacks is currently about 9 percent. The incidence at birth of the disabling sickle-cell anemia is estimated at 2 per 1,000 individuals.

The disappearance of malaria disrupts the balanced polymorphic state, and the gene for the sickle-cell anemia begins to decline in frequency. If the reproductive fitness of the heterozygous carrier is equal to that of the normal homozygote, and the fitness of the sickle-cell anemic individual is less than normal, then the incidence of the heterozygous carrier will become depressed generation after generation. At present, medical researchers are attempting to find a cure for sickle-cell anemia. If sickle-cell anemic individuals could be completely cured, then the selective process would be thwarted and the sickle-cell gene would no longer decline in frequency.

Tay-Sachs Disease

Tay-Sachs disease is a recessively inherited condition that is untreatable and fatal. The basic defect is the abnormal accumulation in the brain cells of a fatty substance (specifically, ganglioside GM_2) as a consequence of the absence of a particular enzyme (hexosaminidase A). The storage of massive amounts of the lipid leaks to profound mental and motor deterioration. Affected children appear normal and healthy at birth, but within six months the nerves of the brain and spinal cord exhibit marked signs of deterioration. Listless and irritable at first, the infant finds it increasingly difficult to sit up or stand. By the age of one year, the child lies helplessly in his crib. He becomes mentally retarded, progressively blind, and finally paralyzed. The disease takes its lethal toll by the age of three to four years. There are no known survivors and no cure.

A feature of special interest is that 9 out of 10 affected children are of Jewish heritage. It is especially common among the Ashkenazi Jews of northeastern European origin, particularly from provinces in Lithuania and Poland. In the United States, Tay-Sachs disease is about 100 times more relevant among the Ashkenazi Jews than among other Jewish (Sephardi) groups and non-Jewish populations. It is estimated that 1 of 40 Jewish persons is heterozygous, whereas only 1 of 380 non-Jewish persons is a heterozygous carrier. If the high incidence of heterozygotes is maintained by mutation alone, then an extraordinarily high mutation rate of the detrimental recessive gene would have to be postulated. However, some reproductive advantage for the heterozygote carrier would seem the most plausible explanation.

It has been suggested that the Ashkenazi Jews, who have lived for many generations in the urban ghettos in Poland and the Baltic states, have been exposed to different selective pressures than other Jewish groups who have lived in countries around the Mediterranean and Near East. The densely populated urban ghettos may have experienced repeated outbreaks of infectious diseases. In 1972, the geneticist Ntinos Myrianthopoulos of the National Institutes of Health in Bethesda, Maryland, presented data that show that pulmonary tuberculosis is virtually absent among grandparents of children afflicted with Tay-Sachs disease, although the incidence of Jewish tuberculosis patients from eastern Europe is relatively high. The findings indicate that the heterozygous carrier of Tay-Sachs disease is resistant to pulmonary tuberculosis in regions where this contagious disease is prevalent.

Cystic Fibrosis

One of the most serious recessive disorders among chidlren is *cystic fibrosis* of the pancreas. About 6,000 infants are born with the disease each year in the United States. There are at least 5 million heterozygous carriers, and 1 child in 2,500 is born with the disease. Virtually all patients with cystic fibrosis are white; blacks and Orientals are rarely afflicted.

Less than 35 years ago, cystic fibrosis was not even recognized as a distinct medical entity. Today it has the unenviable reputation of being one of the most important disorders of childhood. Intestinal obstruction is the first symptom of the disease; the infant's stools are frequent, large, and foul. The pancreas secretes an abnormally thick or viscid mucous material that blocks digestion in the intestinal tract. In addition, all mucus-secreting tissues of affected infants are abnormal. The sticky mucus produced by the lungs is particularly serious; the lungs become clogged and the child has repeated bouts of pneumonia. A characteristic finding, by which cystic fibrosis is readily diagnosed, is the increased amounts of salt in the sweat.

Before the introduction of antibiotics, affected children invariably died in infancy of constant infection. New drugs, inhaled as vapor, soften the thick mucus of the lungs and have enabled some children to weather the difficult first years. The widespread application of antibiotics has led to a considerable reduction in early mortality. In fact, several women patients with cystic fibrosis responded so well to the special treatment that they now have children of their own. All these children are phenotypically normal. But these normal children born of affected mothers harbor the recessive gene for cystic fibrosis and can pass on the defective gene to their offspring.

A reproductive advantage for the heterozygote would account for the high frequency of the recessive gene for cystic fibrosis in white populations. The magnitude of the selective advantage of the heterozygous carrier over the normal homozygote need only be 2 percent to maintain an incidence of the disorder at 1 in 2,500 births. To ascertain the reproductive advantage, one could obtain information on the fertility of heterozygous parents. However, the overall reproductive performance of heterozygous parents tends to be biased by the birth of a child with cystic fibrosis and by the knowledge that this lethal hereditary disorder can arise in subsequent pregnancies.

An unbiased estimate of the fertility of unsuspecting heterozygotes, who unknowingly perpetuate the gene, can be obtained by assessing the fertility of the *grandparents* of the affected children. It may be assumed that at least one of the grandparents (maternal or paternal) of an affected child is a heterozygote. Thus, the size of the sibship from grandparental matings of the type *AA* × *Aa* (where *a* is the allele for cystic fibrosis) can be compared with the sibship size of grandparental matings of the type *AA* × *AA*. It should be noted that family size cannot be attributed to knowledge of the disorder, since by the time a grandchild with cystic fibrosis has been identified, the fertility of the grandparents has been completed. Such a study was undertaken in 1967 by Alfred Knudson and his co-workers at the City of Hope Medical Center at Duarte, California. The grandparents of patients with cystic fibrosis were found to have an average of 4.34 offspring compared with 3.43 offspring for the appropriate control group. Thus, on the average, matings of type *AA* × *Aa* have one additional child compared with *AA* × *AA* matings. The greater reproductive fitness of the heterozygous carrier suggests that the detrimental gene for cystic fibrosis is being maintained at exceptionally high frequencies by balanced polymorphism.

Implications of Balanced Polymorphism

According to the classical concept of selection (discussed in chapter 5), a deleterious gene has a harmful effect when homozygous and virtually no expression in the heterozygous state. Deleterious genes will be reduced in frequency by selection to very low levels, and will be maintained in the population by recurrent mutations. The fittest individuals are homozygous for the normal, or "wild-type," allele at most loci. Stated another way, the fittest individuals carry relatively few deleterious genes in the heterozygous state.

In this chapter, we have considered examples of genes that impair fitness when homozygous and actually improve fitness in the heterozygous state. To what extent are individuals heterozygous at their gene loci? Geneticists estimate that 30 percent of the loci in an individual exist in two or more allelic forms. If the alternative alleles at these polymorphic loci are maintained by selection that favors the heterozygote, then low fitness would have to be assigned to an unusually large number of homozygous loci. In other words, an unreasonably high level of unfitness would prevail in the population because of the selective disadvantage of many alleles in the homozygous state.

The possibility exists that most of the alternative alleles at polymorphic loci are *selectively neutral*—that is, the different alleles at one locus confer neither selective advantage or disadvantage on the individual. Much of the observed variation, then, would represent merely the accumulation of neutral mutations. The notion of selectively neutral alleles has been much debated. Are there mutant alleles whose effects on fitness are not at all different from the more frequent alleles that lead to a normal phenotype? And, can some of these neutral mutant alleles reach frequencies purely by chance as high as the frequencies that characterize the state of balanced polymorphism? These vexing questions have yet to be resolved satisfactorily.

Selected Readings

Allison, A. C. 1956. Sickle cells and evolution. *Scientific American,* August, pp. 87–94.

Anfinsen, C. B. 1963. *The molecular basis of evolution.* New York: John C. Wiley & Sons.

Chernoff, A. 1959. The distribution of the thalassemia gene: a historical review. *Blood* 14:899–912.

Dobzhansky, T. 1962. *Mankind evolving.* New Haven: Yale University Press.

Harris, H. 1969. Enzyme and protein polymorphism in human populations. *British Medical Bulletin* 25:5–13.

Ingram, V. 1963. *The hemoglobins in genetics and evolution.* New York: Columbia University Press.

Knudson, A. G.; Wayne, L.; and Hallet, W. Y. 1967. On the selective advantage of cystic fibrosis heterozygotes. *American Journal of Human Genetics* 19:388–92.

Livingstone, F. B. 1958. Anthropological implications of sickle cell gene distribution in West Africa. *American Anthropologist* 60:533–62.

―――. 1967. *Abnormal hemoglobins in human populations.* Chicago: Aldine Publishing Co.

Myrianthopoulos, N. C. 1972. Founder effect in Tay-Sachs disease unlikely. *American Journal of Human Genetics* 24:341–42.

Myrianthopoulos, N. C., and Aronson, S. M. 1966. Population dynamics of Tay-Sachs disease. I. Reproductive fitness and selection. *American Journal of Human Genetics* 18:313–27.

Price, J. 1967. Human polymorphism. *Journal of Medical Genetics* 4:44–67.

Sheppard, P. M. 1959. *Natural selection and heredity.* New York: Philosophical Library.

Wiesenfeld, S. L. 1967. Sickle-cell trait in human biological and cultural evolution. *Science* 157:1134–40.

8
Genetic Drift and Gene Flow

The peculiar multilegged condition of the bullfrog (see fig. 1.1) is a rarity in nature. Yet, we witnessed in a particular locality in 1958 an exceptionally high incidence of this defective trait. We have surmised that the deformity is caused by a recessive mutant gene. Since harmful recessive genes in a population tend to be carried mostly in the heterozygous state, the multilegged frogs probably arose from matings of heterozygous carriers. The probability that two or more carriers will actually meet is obviously greater in a small population than in a large breeding assemblage. In fact, the number of matings of carriers of a particular recessive gene is a population is mainly a function of the size of the population. Most populations are not infinitely large, and many fluctuate in size from time to time.

During a period when a population is small, chance matings and segregations could lead to an uncommonly high frequency of a given recessive gene. For example, it is not unthinkable that an unduly harsh winter sharply reduced the size of our particular bullfrog population. By sheer chance, an unusually large proportion of heterozygous carriers of the multilegged condition might have survived the winter's severity and prevailed as parents in the ensuing spring's breeding aggregation. In this manner, the "multilegged" gene, although not at all advantageous, would occur with an extraordinarily high incidence in the new generation of offspring. Such a fortuitous change in the genetic makeup of a population that may arise when the population becomes restricted in size is known as *genetic drift*.

Role of Genetic Drift

Examination of a natural situation demonstrating genetic drift will lead us into a simplified mathematical consideration of the concept. Coleman Goin, a naturalist at the University of Florida, studied the distribution of pigment variants of a terrestrial frog, known impressively as *Eleutherodactylus ricordi planirostris,* but commonly as the greenhouse frog. This frog may possess either of two pigmentary patterns, mottled or striped (fig. 8.1). A unique feature of the greenhouse frog is the terrestrial development of its eggs. That is, the eggs need not be submerged completely in water, but can develop in moist earth. This important quality may have considerable bearing on the dispersal of the frogs. Goin reared eggs successfully in a flowerpot two-thirds filled with beach sand and placed it in a finger bowl of water. An examination of a large number of progeny hatched from many different clutches of eggs revealed that the striped pattern is dominant to the mottled pattern.

The greenhouse frog is widespread in Cuba and the Bahama Islands, and has only recently become established in Florida. Cuba apparently has been the center of dispersal from which the Florida populations have been derived. The present distribution of the greenhouse frog in Florida consists of a series of small, isolated colonies. As shown in figure 8.1, the proportions of the two patterns vary in different colonies in Florida. In several colonies, moreover,

---

Figure 8.1   Distribution of the greenhouse frog in Florida, and the relative frequencies of the two pattern variants, *mottled* and *striped.* The populations are small and isolated, and differ appreciably in the incidences of mottled and striped forms. The varied frequencies may not be due to natural selection, but may represent the outcome of chance fluctuations of genes, or genetic drift. (Based on studies by Coleman Goin.)

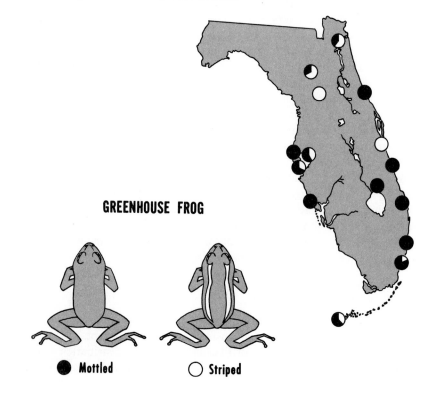

GREENHOUSE FROG

● Mottled   ○ Striped

only the mottled type occurs. What can account for the local preponderance of one or the other pattern, or even the absence of one of the contrasting patterns?

Goin conjectures that the greenhouse frog was introduced into Florida by means of clutches of eggs accidentally included in shipments of plants from Cuba, a distinct possibility in view of the terrestrial development of the eggs. Thus, a single clutch of introduced eggs could initiate a small colony, which, in turn, would establish at the outset a given pattern or proportion of patterns. The presence of only mottled forms in a population may be due to the chance circumstance that only mottled eggs were introduced. Or, perhaps both striped and mottled eggs were included in the shipment, but by sheer accident, one type was lost in succeeding generations.

It should be understood that Goin has not proved that the unusual distribution and frequency of the two pigment patterns are due solely to chance. The demonstration of genetic drift in any natural population is an extremely difficult task. Genetic drift of the variant pigment patterns is, however, a reasonable explanation.

Theory of Genetic Drift

The theory of genetic drift was systematically developed in the 1930s by the geneticist Sewall Wright, then at the University of Chicago and later at the University of Wisconsin. In fact, the phenomenon of drift is frequently called the *Sewall Wright effect*. The Sewall Wright effect refers specifically to the random fluctuations (or drift) of gene frequencies from generation to generation in a population of small size. Because of the limited size of the breeding population, the gene pool of the new generation may not be at all representative of the parental gene pool from which it was drawn.

The essential features of the process of drift may be seen in the following modest mathematical treatment. Let us assume that the numerous isolated colonies in Florida were each settled by only two frogs, a male and a female, both having an *Aa* genotype. Let us further suppose that each mated pair produces only two offspring. The possible genotypes of the progeny, and the chance associations of the genotypes, are shown in table 8.1.

Several meaningful considerations emerge from table 8.1. For example, the chance that the first offspring from a cross of two heterozygous parents will be *AA* is 1/4. The second event is independent of the first; hence, the chance that the second offspring will be *AA* is also 1/4. The chance that *both* offspring will be *AA* is the product of the separate probabilities of the two independent events, $1/4 \times 1/4$, or 1/16.

We may now ask: What is the probability of producing two offspring, one *AA* and the other *Aa, in no particular order?* From table 8.1, we can see that the chance of obtaining an *AA* individual followed by an *Aa* individual is 2/16. Now, the wording of our question requires that we consider a second possibility, that of an *Aa* offspring followed by an *AA* offspring (also 2/16).

Table 8.1
Chance Distribution of Offspring of Two Heterozygous Parents (Aa × Aa)

Genotype of first offspring	Probability of first event	Genotype of second offspring	Probability of second event	Total Probability
AA	1/4	*AA*	1/4	Both offspring *AA*, 1/16
AA	1/4	*Aa*	2/4	*AA* followed by *Aa*, 2/16
AA	1/4	*aa*	1/4	*AA* followed by *aa*, 1/16
Aa	2/4	*AA*	1/4	*Aa* followed by *AA*, 2/16
Aa	2/4	*Aa*	2/4	Both offspring *Aa*, 4/16
Aa	2/4	*aa*	1/4	*Aa* followed by *aa*, 2/16
aa	1/4	*AA*	1/4	*aa* followed by *AA*, 1/16
aa	1/4	*Aa*	2/4	*aa* followed by *Aa*, 2/16
aa	1/4	*aa*	1/4	Both offspring *aa*, 1/16

These two probabilities must be added together to arrive at the chance of producing the two genotypes irrespective of the order of birth. Hence, in the case in question, the chance is 2/16 + 2/16, or 4/16. In like manner, it may be ascertained that the expectation of obtaining one *AA* and one *aa* offspring (in no given order) is 2/16; and that of producing one *Aa* and one *aa* (in any sequence) is 4/16.

The essential point is that any one of the above circumstances may occur in a given colony. We may concentrate on one situation. The probability that a colony will have only two *AA* offspring is 1 in 16. Thus, by the simple play of chance, the parents initiating the colony might not leave an *aa* offspring. The *a* gene would be immediately lost in the population. Subsequent generations descended from the first-generation *AA* individuals would contain, barring mutation, only *AA* types. Chance alone can thus lead to an irreversible situation. A gene once lost could not readily establish itself again in the population. The decisive factor is the size of the population. When populations are small, striking changes can occur from one generation to the next. Some genes may be lost or reduced in frequency by sheer chance; others may be accidentally increased in frequency. Thus, the genetic architecture of a small population may change irrespective of the selective advantage or disadvantage of a trait. Indeed, a beneficial gene may be lost in a small population before natural selection has had the opportunity to act on it favorably.

Founder Effect

When a few individuals or a small group migrate from a main population, only a limited portion of the parental gene pool is carried away. In the small migrant group, some genes may be absent or occur in such low frequency that they may be easily lost. The unique frequencies of genes that arise in populations derived from small bands of colonizers, or "founders," has been called the *founder effect*. This expression essentially emphasizes the conditions or circumstances that foster the operation of genetic drift.

The American Indians afford a possible example of the loss of genes by the founder principle. North American Indian tribes, for the most part, surprisingly lack the gene (designated I^B) that governs type B blood. However, in Asia, the ancestral home of the American Indian, the I^B gene is widespread. The ancestral population of Mongoloids that migrated across the Bering Strait to North America might well have been very small. Accordingly, the possibility exists that none of the prehistoric immigrants happened to be of blood group B. It is also conceivable that a few individuals of the migrant band did carry the I^B gene but they failed to leave descendants.

The interpretation based on genetic drift should not be considered as definitive. The operation of natural selection cannot be flatly dismissed. Most of the North American Indians possess only blood group O, or stated another way, contain only the blood allele i. With few exceptions, the North American Indian tribes have lost not only blood group allele I^B but also the allele that controls type A blood (I^A). The loss of both alleles, I^A and I^B, by sheer chance perhaps defies credibility. Indeed, many modern students of evolution are convinced that some strong selective force led to the rapid elimination of the I^A and I^B genes in the American Indian populations. If this is true, it would offer an impressive example of the action of natural selection in modifying the frequencies of genes in a population.

North American Indians are also known to have a high frequency of albinism. The incidence of albinism among the Cuna Indians of the San Blas Province in Panama is about 1 in 200, which contrasts sharply with the 1 in 20,000 figure for European Caucasians. The Hopi Indians of Arizona and the Zuñi Indians of New Mexico, like the Cuna Indians, are also remarkable in their high numbers of albino individuals.

Since these American Indian populations are small, one might suspect the operation of genetic drift, notwithstanding the deleterious effects of the albino gene. It is difficult to imagine, however, that this detrimental gene could reach a high frequency on the basis of chance alone independently in several American Indian populations. Charles M. Woolf, the geneticist at Arizona State University, has suggested that the high incidence of albinism among the Hopi Indians of Arizona reflects an inimitable form of selection that he terms *cultural selection*. Albinos have been highly regarded in the traditional Hopi society, and actually have enjoyed appreciable success in regard to sexual activity. The albino male has been admired, and some have become legendary for leaving large numbers of offspring. Woolf further notes that the fading of old customs among the Hopi Indians is beginning to nullify any reproductive advantage held by albino males in past generations. The frequency of albinism may be expected to decline with the dissolution of the traditional Hopi way of life.

Religious Isolates

The most likely situation to witness the phenomenon of genetic drift is one in which the population is virtually a small, self-contained breeding unit, or *isolate,* in the midst of a larger population. This typifies the Dunkers, a very small religious sect in eastern Pennsylvania. The Dunkers are descendants of the Old German Baptist Brethren, who came to the United States in the early eighteenth century. Bentley Glass, then at Johns Hopkins University, studied the community of Dunkers in Franklin County, Pennsylvania, which numbers about 300 individuals. In each generation, the number of parents has remained stable at about 90. The Dunkers live on farms intermingled with the general population, but are genetically isolated by rigid marriage customs. The choice of mates is restricted to members within the religious group.

Glass, with his colleagues, compared the frequencies of certain traits from the Dunker community, the surrounding heterogeneous American population, and the population in western Germany from which the Dunker sect had emigrated two centuries ago. Such a comparison of a small isolate with its large host and parent populations should reveal the effectiveness, if any, of genetic drift. In other words, if the small isolated population shows aberrant gene frequencies as compared to the large parent population, and if the other forces of evolution can be excluded, then the genetic differences can be ascribed to drift.

Analyses were made of the patterns of inheritance of three blood group systems—the ABO blood groups, the MN blood types, and the Rh blood types. In addition, data were accumulated on the incidences of four external traits—namely, the configuration of the ear lobes (which either may be attached to the side of the head or hang free), right- or left-handedness, the presence or absence of hair on the middle segments of the fingers (mid-digital hair), and "hitch-hiker's thumb," technically termed distal hyperextensibility (fig. 8.2).

The frequencies of many of these traits are strikingly different in the Dunker community from those of the general United States and West German populations. Blood group A is much more frequent among the Dunkers; the O group is somewhat rarer in the Dunkers; and the frequencies of groups B and AB have dropped to exceptionally low levels in the Dunker community. In fact, the I^B gene had almost been lost in the isolate. Most of the carriers of the I^B gene were not born in the community, but were converts who entered the isolate by marriage.

A noticeable change has occurred also in the incidences of the M and N blood types in the Dunker community. Type M has increased in frequency, and type N has dwindled in frequency as compared with the incidences of these blood types in either the general United States population or the West German population. Only in the Rh blood groups do the Dunkers conform closely to their surrounding large population.

Figure 8.2 Inheritable physical traits—nature of ear lobes, "hitchhiker's thumb," mid-digital hair, and handedness—studied by Bentley Glass and his co-workers in members of the small religious community of Dunkers in Pennsylvania. The distinctive frequencies of most of these traits in the Dunker population suggests the operation of genetic drift.

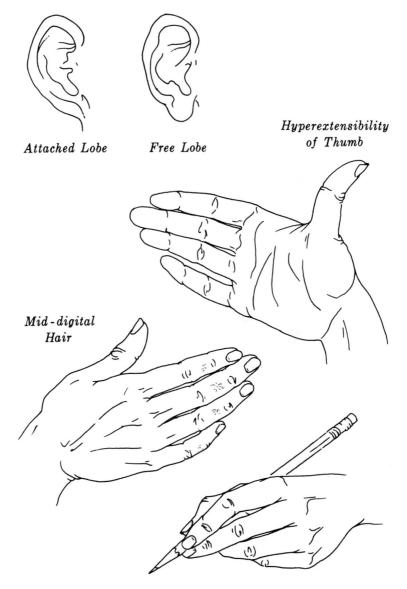

Attached Lobe   *Free Lobe*   *Hyperextensibility of Thumb*

Mid-digital Hair

Left-handedness

In physical traits, equally striking differences were found. Briefly, the frequencies of mid-digital hair patterns, distal hyperextensibility of the thumb, and attached ear lobes are significantly lower in the Dunker isolate than in the surrounding American populations. The Dunkers do, however, agree well with other large populations in the incidence of left-handedness. It would thus appear that the peculiar constellation of gene frequencies in the Dunker community—some uncommonly high, others uniquely low, and still others, unchanged from the general large population—can be best attributed to chance fluctuations, or genetic drift.

There is no concurrence of opinion among evolutionists concerning the operation of genetic drift in natural populations, but few would deny that small religious isolates have felt the effect of random sampling. It should be clear, however, that genetic drift becomes ineffectual when a small community increases in size. Fluctuations or shifts in gene frequencies in large populations are determined almost exclusively by selection.

Amish of Pennsylvania

We have seen that gene frequencies in small religious isolates may differ significantly from the original large populations from which the isolates were derived. Another feature of small isolates is the occurrence of rare recessive traits in greater numbers than would be expected from random mating in a large population. This is witnessed among the Old Order Amish societies in the eastern United States.

The Amish sect is an offshoot of the Mennonite Church; both religious groups settled in the United States to escape persecution in Europe in past centuries. The Amish are old-fashioned, rural-living people who cultivate the religious life apart from the world. Present-day communities were founded by waves of Amish immigration that began about 1720 and continued until about 1850. The vast majority of Amish live in relatively isolated colonies in Pennsylvania, Ohio, and Indiana. Each community is descended from a small immigrant stock, as attested by the relatively few family names in a given community. Analyses by the geneticist Victor A. McKusick of Johns Hopkins University have shown that eight names account for 80 percent of the Amish families in Lancaster County, Pennsylvania. Other Amish communities also are characterized by a high frequency of certain family names, as table 8.2 shows.

Marriages have been largely confined within members of the Amish sect, with a resulting high degree of consanguinity. Marriages of close relatives have tended to promote the meeting of two normal, but carrier, parents. Four recessive disorders manifest themselves with uncommonly high frequencies, each in a different Amish group: the Ellis-van Creveld syndrome, pyruvate-kinase-deficient hemolytic anemia, Hemophilia B (Christmas disease), and a form of limb-girdle muscular dystrophy (Troyer syndrome).

Table 8.2
Old Order Amish Family Names in Three American Communities*

Lancaster Co., Pa.		Holmes Co., Ohio		Mifflin Co., Pa.	
Stolzfus**	23%	Miller	26%	Yoder	28%
King	12%	Yoder	17%	Peachey	19%
Fischer	12%	Troyer	11%	Hostetler	13%
Beiler	12%	Hershberger	5%	Byler	6%
Lapp	7%	Raber	5%	Zook	6%
Zook	6%	Schlabach	5%	Speicher	5%
Esh***	6%	Weaver	4%	Kanagy	4%
Glick	3%	Mast	4%	Swarey	4%
	81%		77%		85%

Totals:

1,106 families, 1957	1,611 families, 1960	238 families, 1951

* From data compiled by Victor A. McKusick of Johns Hopkins University.
** Including Stolzfoos.
*** Including Esch.

We may consider in some detail the Ellis-van Creveld syndrome, which occurs in the Lancaster County population (see fig. 1.5). Fifty-two affected persons have been identified in 30 sibships, most of which have unaffected parents. Pedigree analysis has revealed that Samuel King and his wife, who immigrated in 1744, are ancestral to all parents of the sibships. Either Samuel King or his wife carried the recessive gene. None of their children were affected, but subsequent generations were. Evidently, previously concealed detrimental recessive genes are brought to light by the increased chances of two heterozygotes meeting in a small population.

Consanguinity and Genetic Drift

Small populations not only provide opportunities for genetic drift but for consanguinity as well. As witnessed among the Amish, offspring afflicted with recessive disorders arise more often from unions of close relatives than from marriages of unrelated persons. Yet the Amish sternly frown on first-cousin marriages. Nonetheless, although consanguineous marriages are not made by choice, the limited size of the Amish population restricts the availability of potential mates and virtually forces marriages of close relatives. Indeed, second-cousin and third-cousin marriages are rather common among the Amish. The relationship between genetic drift and inbreeding thus becomes clear: the closed nature of a small population creates a situation wherein few nonrelatives are present in the population. Because matings of genetically related individuals increase the probability of homozygosity in the progeny, a relatively large number of rare recessive homozygotes are expected, and do occur.

The smaller the population and the longer it has been isolated, the greater the chance that most members of the population are related to each other through common ancestors. This is exemplified by the isolated island of Tristan da Cunha in the South Atlantic. This small island is of historical interest. In 1816, a British military garrison was established on Tristan for the sole purpose of safeguarding against the escape of Napoleon from the neighboring, though distant, St. Helena. Within a few months, it became glaringly apparent that Tristan was inconsequential to Napoleon's safekeeping. The battery of soldiers was hastily withdrawn, but a Scots corporal, William Glass, his family, and a few soldiers remained behind. From 1817 until 1908, women from St. Helena and shipwrecked sailors joined the community to bring the total to a mere 15 individuals. The present population of nearly 270 traces their origin to the 15 early settlers.

The English geneticist D. F. Roberts has calculated that in Tristan today the probability is high that any two young individuals contemplating marriage share numerous genes due to the common ancestry of their parents. Roberts stresses that the relatively high level of consanguinity on the island does not reflect a conscious preference for marriage between close relatives. Rather, random mating prevails, but the marriages prove to be consanguineous because most of the potential mates are already relatives by common descent.

Gene Flow

A rich archeological record reveals appreciable movement on the part of early human populations. Some migrations were sporadic, in small groups; others were more or less continual streams involving large numbers of peoples. Large-scale immigrations followed by interbreeding have the effect of introducing new genes to the host populations. The diffusion of genes into populations through migrations is referred to as *gene flow*.

The graded distribution of the I^B blood-group gene in Europe represents the historical consequence of invasions by Mongolians who pushed westward repeatedly between the sixth and sixteenth centuries (fig. 8.3). There is a high frequency of the I^B gene in central Asia. In Europe, the frequency of the I^B gene diminishes steadily from the borders of Asia to a low level of 5 percent or less in parts of Holland, France, Spain, and Portugal. The Basque peoples, who inhabit the region of the Pyrenees in Spain and France, have the lowest frequency of the I^B gene in Europe—below 3 percent. From a biological standpoint, the Basque community of long standing is a cohesive, endogamous mating unit. The exceptionally low incidence of the I^B gene among the Basques may be taken to indicate that there has been little intermarriage with surrounding populations. It is possible that a few centuries ago the I^B gene was completely absent from the self-contained Basque community.

The Basques are also of interest in having a high frequency of the Rh-negative allele, r. Individuals who carry two recessive genes (rr) are said to be Rh-negative. An Rh-positive person has the dominant gene, either in the homozygous state (RR) or the heterozygous state (Rr). The Rh alleles are associated with a blood disease of the fetus and newborn, known variously as

Figure 8.3 Gradient of frequencies of the I^B blood-group gene from central
Asia to western Europe. (Based on studies by A. E. Mourant.)

erythroblastosis fetalis, hemolytic disease of the newborn, or simply *Rh disease.* An infant afflicted with Rh disease is always the heterozygote (*Rr*), born to Rh-negative mothers and Rh-positive fathers.

The Basques are descendants of an early European population that was almost exclusively *rr.* The Rh-positive allele (*R*) was probably introduced by migrants into the Basque community, much in the same way that the Rh-negative allele (*r*) is presently being introduced in populations in China. Until recently, Rh disease was virtually unknown in China. All Chinese women are Rh-positive (*RR*). However, intermarriage between immigrant Americans and the native Chinese has led to the introduction of the Rh-negative gene (*r*) in the Chinese population. No Rh disease would be witnessed in the immediate offspring of American men and Chinese women. Conversely, all marriages of Rh-negative American women (*rr*) and Rh-positive men (*RR*) would be of the incompatible type. All children by Chinese fathers would be Rh-positive (*Rr*) and potential victims of hemolytic disease.

Whereas American immigrants introduce the Rh-negative gene (*r*) into Chinese populations where it formerly was not present, Chinese immigrants (all of whom are *RR*) introduce more Rh-positive genes (*R*) into the American populations, thus diluting the Rh-negative gene pool in the United States. Initially, the Rh-positive Chinese men (*RR*) married to Rh-negative American women (*rr*) would result in an increased incidence of Rh-diseased infants. In later generations, however, the frequency of Rh-negative women in the United

States would be lower, inasmuch as women of mixed Chinese-American origin would be either *RR* or *Rr*, predominantly the former. Thus, in the United States, the long-range effect of Chinese-American intermarriage is a reduction in the incidence of hemolytic disease of the newborn.

White Genes in American Blacks

In the United States, from a beginning more than 250 years ago, there has occurred an admixture of whites with the blacks who were brought from Africa to the New World as slaves. The overall white contribution to the American black gene pool has been estimated at approximately 25 percent. This percentage must be considered as a gross estimate, inasmuch as different black populations in the United States have undergone varied degrees of admixture.

The geneticist T. Edward Reed of the University of Minnesota has measured the frequency with which a particular blood-group gene appears in American whites, African blacks, and American blacks. The specific gene is the Fy^a gene (the Duffy factor) of the Duffy blood group system. As seen in table 8.3, the Duffy factor is virtually absent in the African black populations

Table 8.3
Duffy Blood Factor (Fy^a Gene) in American White, African Black, and American Black Populations

Region	Frequency of Duffy Factor	Percentage of White Contribution
A. African Black		
Liberia	0.00	—
Ivory Coast	0.04	—
Upper Volta	0.00	—
Dahomey	0.00	—
Ghana (Accra)	0.00	—
Nigeria (Lagos)	0.00	—
B. American White		
Oakland, California	0.43	
Evans and Bullock Counties, Georgia	0.42	
C. American Black		
1. Non-Southern		
New York, New York	0.08	18.9
Detroit, Michigan	0.11	26.0
Oakland, California	0.09	22.0
2. Southern		
Charleston, South Carolina	0.02	3.7
Evans and Bullock Counties, Georgia	0.05	10.6

Based on data compiled by T. Edward Reed of the University of Minnesota.

from which most of the original immigrant slaves were derived. The highest frequency of the Duffy factor in the stem populations in Africa is an exceedingly low 0.04, whereas the value for representative white populations in the United States is 0.4, or ten times as great (table 8.3). Among blacks residing in New York City, Detroit, and Oakland, the gene reaches a frequency of about 0.10. This signifies that the magnitude of white ancestry in blacks in these localities may be as high as 26 percent. In contrast, the frequency of the Duffy factor among blacks in Charleston, South Carolina, is 0.02, which indicates that the black population in Charleston has only a small amount of white ancestry. Thus, although the extent of white-black hybridization has been variable, it is clear that various American black populations derive between 4 percent and 26 percent of their genes from white ancestors, and these white genes have been introduced through hybridization since 1700.

Selected Readings

Boyd, W. C. 1963. Four achievements of the genetical method in physical anthropology. *American Anthropologist* 65:243–52.

Candela, P. B. 1942. The introduction of blood-group B into Europe. *Human Biology* 14:413–43.

Cavalli-Sforza, L. L. 1969. "Genetic drift" in an Italian population. *Scientific American,* August, pp. 30–37.

Cavalli-Sforza, L. L.; Barrai, I.; and Edwards, A. W. F. 1964. Analysis of human evolution under random genetic drift. *Cold Spring Harbor Symposium on Quantitative Biology* 29:9–20.

Dunn, L. C. 1966. *Heredity and evolution in human populations.* New York: Atheneum Publishers.

Dunn, L. C., and Dunn, S. P. 1957. The Jewish community of Rome. *Scientific American,* March, pp. 118–28.

Glass, B. 1953. The genetics of the Dunkers. *Scientific American,* August, pp. 76–81.

―――. 1954. Genetic changes in human populations, especially those due to gene flow and genetic drift. *Advances in Genetics* 6:95–139.

―――. 1956. On the evidence of random genetic drift in human populations. *American Journal of Physical Anthropology* 14:541–55.

Glass, B., and Li, C. C. 1953. The dynamics of racial intermixture—an analysis based on the American Negro. *American Journal of Human Genetics* 5:1–20.

Goldsby, R. A. 1971. *Race and races.* New York: Macmillan Co.

Goldschmidt, E. 1963. *The genetics of migrant and isolate populations.* Baltimore: Williams & Wilkins Co.

McKusick, V. A.; Hostetler, J. A.; Egeland, J. A.; and Eldridge, R. 1964. The distribution of certain genes in the Old Order Amish. *Cold Spring Harbor Symposium on Quantitative Biology* 29:99–114.

Mourant, A. E. 1954. *The distribution of the human blood groups.* Oxford. England: Blackwell Scientific Publications.

Pollitzer, W. S. 1958. The Negroes of Charleston (S.C.): a study of hemoglobin types, serology and morphology. *American Journal of Physical Anthropology* 16:241–63.

Reed, T. E. 1969. Caucasian genes in American Negroes. *Science* 165:762–68.

Woolf, C. M., and Dukepoo, F. C. 1969. Hopi Indians, inbreeding, and albinism. *Science* 164:30–37.

Wright, S. 1951. Fisher and Ford on "The Sewall Wright Effect." *American Scientist* 39:452–58.

9
Races and Species

Any large assemblage of a particular organism is generally not distributed equally nor uniformly throughout its territory or range in nature. A widespread group of plants or animals is typically subdivided into numerous local populations, each physically separated from the others to some extent. The environmental conditions in different parts of the range of an organism are not likely to be identical. We may thus expect that a given local population will consist of genetic types adapted to a specific set of prevailing environmental conditions. The degree to which each population maintains its genetic distinctness is governed by the extent to which *interbreeding* between the populations occurs. A free interchange of genes between populations tends to blur the differences between the populations. But what are the consequences when gene exchange between populations is greatly restricted or prevented? This chapter addresses itself to this question.

Variation Between Populations

Our first consideration is to demonstrate that inheritable variations exist among the various breeding populations in different geographical localities of an organism. Jens Clausen, David Keck, and William Hiesey of the Carnegie Institution of Washington at Stanford, California, have shown that each of the populations of the yarrow plant, *Achillea lanulosa,* from different parts of California is adapted to its respective habitat. As figure 9.1 shows, the variations in height of the plant are correlated with altitudinal differences. The shortest plants are from the highest altitudes, and the plants increase in height in a gradient fashion with decreasing altitude. The term *cline,* or character gradient, has been applied to such situations where a character varies more or less continuously with a gradual change in the environmental terrain.

Figure 9.1 **Clinal variation** in the yarrow plant, *Achillea lanulosa.* The increase in height of the plant is more or less continuous with decreasing altitude. The plants shown here are representatives from different populations in the Sierra Nevada Mountains of California that were grown in a uniform garden at Stanford, California. Each plant illustrated is one of average height for the given population; the graph adjacent to the plant reveals the distribution of heights within the population. (From Clausen, Keck, and Hiesey, *Carnegie Institution of Washington Publication 581,* 1948.)

The observation by itself that the yarrow plants are phenotypically dissimilar at different elevations does not indicate that they are genetically different. If the observed variations are claimed as local adaptations resulting from natural selection, then a hereditary basis for the differences in height should be demonstrated. It is often difficult to obtain data that disclose the hereditary nature of population differences. In this respect, the studies of Clausen and his co-workers are commendable. The plants shown in figure 9.1 had actually been grown together in a uniform experimental garden at Stanford, California. The plants, transplanted from various localities, developed differently from one another in the same experimental garden, revealing that each population had evolved its own distinctive complex of genes.

Races

The variation pattern in organisms may be discrete, or discontinuous, particularly when the populations are separated from each other by pronounced physical barriers. This is exemplified by the varieties of the carpenter bee (*Xylocopa nobilis*) in the Celebes and neighboring islands of Indonesia (fig. 9.2).

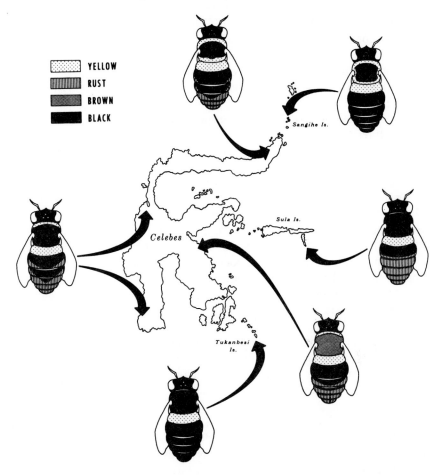

Figure 9.2 Geographic variation of color patterns in females of the carpenter bee, *Mylocopa nobilis,* in the Celebes and neighboring islands in Indonesia. Each geographic race has evolved a distinctive constellation of colors. (Based on studies by J. van der Vecht.)

As shown by the studies of J. van der Vecht of the Museum of Natural History at Leiden in the Netherlands, there are three different varieties on the mainland of Celebes, and at least three kinds on the adjacent small islands. These geographical variants differ conspicuously in the coloration of the small, soft hairs that cover the surface of the body. The first abdominal segment is invariably clothed with bright yellow hairs. However, each variety has evolved a unique constellation of colors on the other abdominal segments and also on the thorax.

The variations in the carpenter bees within and between islands are well defined and easily distinguishable. One may refer to populations with well-marked discontinuities as *races.* Races are simply geographical aggregates of populations that differ in the incidence of genetic traits. How genetically different two assemblages of populations must be to warrant racial designations is an open question.

Some of the problems inherent in delimiting races are exemplified by the different temperature-adapted populations of the North American leopard frog, *Rana pipiens*. John A. Moore, then at Columbia University and later at the University of California, tested the effects of temperature on the development of the embryos of frogs from widely different localities. He wished to ascertain the limits of temperatures that the embryos can endure or tolerate. The findings on four different geographic populations in the eastern United States are shown in figure 9.3.

Embryos of northern *Rana pipiens* populations are most resistant to low temperatures and less tolerant of high temperatures than are embryos from southern populations. Embryos of populations from Vermont and New Jersey have comparable ranges of temperature tolerances. These northern embryos can resist temperatures as low as 5°C. Embryos from Florida differ markedly from those of northern populations. Embryos from southern Florida (latitude 27°N) can tolerate temperatures as high as 35°C, but are very susceptible to low temperatures. Hence, northern and southern populations have become adapted to different environments in their respective territories.

We may refer to the northern populations as the cold-adapted race of the leopard frog, and designate the southern populations as the warm-adapted race. It is evident that we are being arbitrary in drawing a fine line of demarcation between northern and southern races. Data are presently lacking for the geographically intermediate populations, but further studies will probably reveal that the temperature adaptations of the frog embryos change gradually from north to south. Even with the present information, one may wish to recognize more than two temperature-adapted races, or perhaps, as some investigators firmly argue, refrain completely from making racial designations.

Races may be best thought of as units of organization below the species level. In other words, races may be considered as stages in the transformation of populations into species. But what constitutes a species? Up to this point we have assiduously avoided the use of the term *species*. A discussion of the process leading to the formation of species will facilitate understanding of the term itself.

Formation of Species

Let us imagine a large assemblage of land snails subdivided in three geographical aggregations or races, *A, B,* and *C,* each adapted to local environmental conditions (fig. 9.4). There are initially no gross barriers separating the populations from each other, and where *A* meets *B* and *B* meets *C,* interbreeding occurs. Zones of intermediate individuals are thus established between the races, and the width of these zones depends on the extent to which

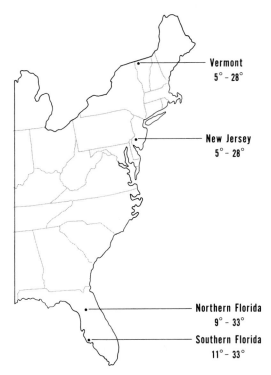

Figure 9.3 Limits of temperature tolerance of embryos of the leopard frog (*Rana pipiens*) from different geographical populations. Embryos of northern populations are more resistant to low temperatures and less tolerant of high temperatures than are embryos from southern populations. (Based on studies by John A. Moore.)

Vermont
5° - 28°

New Jersey
5° - 28°

Northern Florida
9° - 33°

Southern Florida
11° - 33°

the respective populations intermingle. It is important to realize that races are fully capable of exchanging genes with one another.

We may now visualize (fig. 9.4) some striking physical feature, such as a great river, forging its way through the territory and effectively isolating the land snails of race *C* from those of *B*. These two assemblages may be spatially separated from each other for an indefinitely long period of time, affording an opportunity for race *C* to pursue its own independent evolutionary course. Two populations that are geographically separated, like *B* and *C* in our pictorial model, are said to be *allopatric*. (Technically speaking, *A* and *B* are also allopatric, since they, for the most part, occupy different geographical areas.)

After eons of time, the river may dry up and the hollow bed may eventually become filled in with land. Now, if the members of populations *B* and *C* were to extend their ranges and meet again, one of two things might happen. The snails of the two populations may freely interbreed and establish once again a zone of intermediate individuals. On the other hand, the two populations may no longer be able to interchange genes. If the two assemblages

Figure 9.4   Model for the process of geographic speciation. Members of population (or race) C had diverged genetically during geographical isolation in ways that have made them reproductively incompatible with race B when they met again. Race C has thus transformed into a new species, C'.

Geographical Variation

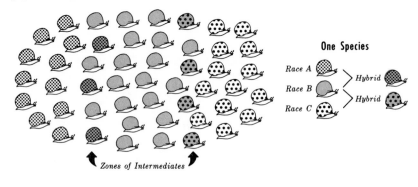

Isolating Action of Geographic Barrier

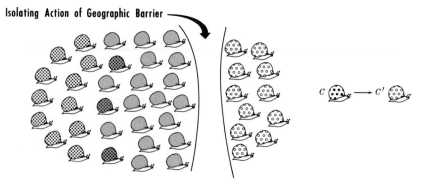

Removal of Geographic Barrier

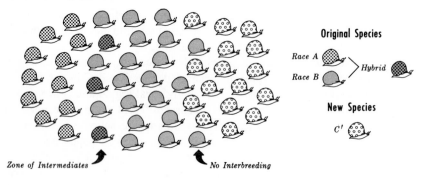

can exist side by side without interbreeding, then the two groups have reached the evolutionary status of separate species. *A species is a breeding community that preserves its genetic identity by its inability to exchange genes with other such breeding communities.* In our pictorial model (fig. 9.4), race *C* has become transformed into a new species, C'. Two species (*A-B* and C') have now arisen where formerly only one existed. It should be noted that races *A* and *B* are treated as members of a single species since no barriers to gene exchange exist between them.

Nomenclature

The scientific names that the taxonomist would apply to our populations of land snails deserve special comment. The technical name of a species consists of two words, in Latin or in latinized form. An acceptable designation of the original species of land snails depicted in figure 9.4 would be *Helix typicus.* The first word is the name of a comprehensive group, the genus, to which land snails belong; the second word is a name unique to the species. The taxonomist would be obliged to create a different latinized second name for the newly derived species of land snail, the C' population in figure 9.4. This new species might well be called *Helix varians.* The name of the genus remains the same since the two species are closely related. The genus, therefore, denotes a group of interrelated species.

The binomial ("two-named") system of nomenclature, universally accepted, was devised by the Swedish naturalist Carolus Linnaeus (stylized later to Carl von Linné) in his monumental work, *Systema Naturae,* first published in 1735. Convention dictates that the first letter of the generic name be capitalized and that the specific name begin with a small letter. It is also customary to print the scientific name of a species in italics, or in a type that is different from that of the accompanying text. A modern refinement of the Linnaean system is the introduction of a third italicized name, which signifies the subspecies. Geographical races are recognized taxonomically as subspecies. Thus, it would be appropriate to designate races *A* and *B* (fig. 9.4) as *Helix typicus elegans* and *Helix typicus eminens,* respectively. Such a species composed of two (or more) subspecies is said to be *polytypic.* A monotypic species is one that is not differentiated into two or more geographical races or subspecies. *Helix varians* would be a monotypic species.

Reproductive Isolating Mechanisms

We have seen that two populations (or races), while spatially separated from each other, may accumulate sufficient genetic differences in isolation to prevent an interchange of genes if they came into contact with one another. When the geographical barrier persists, it is difficult to judge the extent of genetic divergence between the two allopatric populations. Only when the two populations come together again does it become apparent whether or not they have changed in ways that would make them reproductively incompatible. Two populations that come to occupy the same territory are called *sympatric*. The ways or agencies that prevent interbreeding between sympatric species are known as *reproductive isolating mechanisms*.

Reproductive isolating mechanisms are varied forms, and one or more of the different types may be found separating two species. The various types may be grouped into two broad categories. One category includes the prezygotic mechanisms, which serve to prevent the formation of hybrid zygotes. The other category encompasses the postzygotic mechanisms, which act to reduce the viability or fertility of hybrid zygotes. The specific types of isolating mechanisms under these two groupings can be listed as follows:

A. Prezygotic mechanisms
   1. Habitat (ecological) isolation
   2. Seasonal (temporal) isolation
   3. Sexual (ethological) isolation
   4. Mechanical isolation
   5. Gametic isolation
B. Postzygotic mechanisms
   1. Hybrid inviability
   2. Hybrid sterility
   3. Hybrid breakdown

Two related species may live in the same general area, but differ in their ecological requirements. The scarlet oak (*Quercus coccinea*) of eastern North America grows in moist or swampy soils, whereas the black oak (*Quercus velutina*) is adapted to drier soils (fig. 9.5). The two kinds of oaks are thus effectively separated by different *ecological* or *habitat* preferences. Two sympatric species may also retain their distinctness by breeding at different times of the year (*seasonal isolation*). Evidently, cross-fertilization is not feasible between two species of frogs that release their gametes on different months even in the same pond, or between two species of pine that shed their pollen in different periods. The breeding seasons of two species may overlap, but interbreeding may not occur because of the lack of mutual attraction between the sexes of the two species (*sexual isolation*). Among birds, for example, elaborate courtship rituals play important roles in species recognition and the avoidance of interspecific matings.

In many insects, interbreeding of species is hindered by differences in the structure of the reproductive apparatus (*mechanical isolation*). Copulation is not possible because the genitalia of one species is physically incompatible with the genitalia of the other species. In fact, several closely related species of insects can often be accurately classified by their distinctive genitalia. In some instances, the male of one species may inseminate the female of the other species, but the sperm cells may be inviable in the reproductive tract of the female. This form of *gametic isolation* is not unique to animals; in plants, such

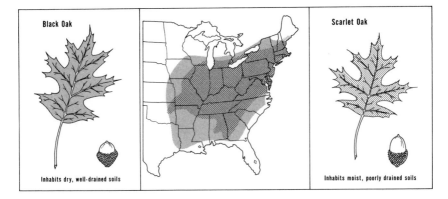

Figure 9.5 **Ecological isolation** is exemplified by the different habitat requirements of two species of oaks. Although both species occur together in eastern United States, the scarlet oak (*Quercus coccinea*) is adapted to the moist bottom lands, whereas the black oak (*Quercus velutina*) is adapted to the dry upland soils.

Black Oak

Inhabits dry, well-drained soils

Scarlet Oak

Inhabits moist, poorly drained soils

as the Jimson weed (*Datura*), the sperm-bearing pollen tube of one species encounters a hostile environment in the flower tissue of the other species and is unable to reach the egg.

Cross-fertilizations between two species may be successful, but the hybrid embryos may be abnormal or fail to reach sexual maturity (*hybrid inviability*). For example, two species of the chicory plant, *Crepis tectorum* and *Crepis capillaris*, can be crossed, but the hybrid seedlings die in early development. Crosses between the bullfrog, *Rana catesbeiana*, and the green frog, *Rana clamitans*, result in inviable embryos. In certain hybrid crosses, such as between females of the toad species *Bufo fowleri* and males of *Bufo valliceps*, the hybrids may survive but are completely sterile (*hybrid sterility*). The familiar example of hybrid sterility is the mule, the offspring of a male ass and a female horse. In some situations, the F_1 hybrids appear to be vigorous and fertile, but the viability of a subsequent generation is very reduced. Such a case of *hybrid breakdown* has been described in several species of cotton— *Gossypium hirsutum* and *Gossypium barbadense*. They produce normal, fertile F_1 hybrids, but the majority of the F_2 hybrid cotton seedlings fail to germinate.

In essence, two populations can remain genetically distinct, and be designated as species, when gene exchange between them is prevented or limited by one or more reproductive isolating mechanisms. More often than not, we are unable to obtain direct evidence for the presence or absence of interbreeding in nature between two groups. The degree of reproductive isolation is then indirectly gauged by the extent to which the members of two populations differ in morphological, physiological, and behavioral characteristics. Two populations that are morphologically very dissimilar are likely to be distinct species. It should be understood, however, that the level of morphological differentiation cannot be used with implicit confidence as a criterion of a species. For example, two species of fruit flies, *Drosophila pseudoobscura* and *Drosophila persimilis,* although reproductively isolated, are almost indistinguishable on morphological grounds.

Origin of Isolating Mechanisms

How do reproductive isolating mechanisms arise? In the 1940s, John A. Moore undertook a series of instructive evolutionary studies on the leopard frog, *Rana pipiens.* The leopard frog is widely distributed in North America, ranging from northern Canada through the United States and Mexico into the lower reaches of Central America. Moore obtained leopard frogs from different geographical populations and crossed them in the laboratory. When frogs from northeastern United States (Vermont) were crossed with their southerly distributed lowland relatives in eastern Mexico (Axtla in San Luis Potosi), the hybrid embryos failed to develop normally. Thus, the geographically extreme members of this species have diverged genetically to the extent that they cannot produce viable hybrids in the laboratory. Indeed, current investigators now formally recognize the existence of several distinct species within the wide-ranging *Rana pipiens* assemblage.

It must be admitted that the possibility of a Vermont frog crossing with a Mexican frog in nature is extremely remote. It took a biologist to bring these two frogs together. Yet, this emphasizes the point that an isolating mechanism, such as hybrid inviability, does not develop for the effect itself; it is simply the natural consequence of sufficient genetic differences having accumulated in two populations during a long geographical separation. The late Hermann J. Muller of Indiana University was among the first to advocate that isolating mechanisms originate as a by-product by genetic divergence of allopatric populations. The genetic changes that arise to adapt one population to particular environmental conditions may also be instrumental in reproductively isolating that population from other populations that are themselves developing adaptive gene complexes. Indeed, the embryos of Vermont leopard frogs differ considerably in their range of temperature tolerance from embryos of eastern Mexican frogs. It might well be, then, that the embryonic defects in hybrids between these northern and southern frogs are associated with the different temperature adaptations of the parental eggs.

If the Vermont and Mexican leopard frogs were ever to meet in nature, then any intercrosses between them would lead to the formation of inviable hybrids. This would represent a wastage of reproductive potential of the parental frogs. Theodosius Dobzhansky has advanced the interesting hypothesis that under such conditions, natural selection would promote the establishment of isolating mechanisms that would guard against the production of abnormal hybrids. In frogs, a normal mating or a mismating in a mixed population depends principally on the discrimination of the female. The productive potential is obviously lower for an undiscriminating female than for a female who leaves normal offspring. If the tendency to mismate is inheritable, then the genes responsible for this tendency will eventually be lost or sharply reduced in frequency by elimination of the indiscriminate females, an elimination effectively accomplished by the inviability of their offspring. Thus, the continual propagation of females that most resist the attentions of "foreign" males will lead eventually to a situation in which mismatings do not occur and abnormal hybrids are no longer produced.

Karl Koopman, an able student of Theodosius Dobzhansky, has tested the thesis that natural selection tends to strengthen, or make complete, the reproductive isolation between two species coexisting in the same territory. Koopman used for experimentation two species of fruit flies, *Drosophila pseudoobscura* and *Drosophila persimilis*. In nature, sexual selection between these two sympatric species is strong, and interspecific matings do not occur. However, in a mixed population in the laboratory, particularly at low temperatures, mismatings do take place. Koopman accordingly brought together members of both species in an experimental cage and purposely kept the cage at a low temperature (16°C). Hybrid flies were produced and were viable, but Koopman in effect made them inviable by painstakingly removing them from the breeding cage when each new generation emerged. Over a period of several generations the production of hybrid flies dwindled markedly and mismatings in the population cage were substantially curtailed. This is a dramatic demonstration of the efficacy of selection in strengthening reproductive isolation between two sympatric species.

Humans: A Single Variable Species

There is only one present-day human species, *Homo sapiens*. Different populations of man can interbreed successfully and, in fact, do. The extensive commingling of populations renders it difficult, if not impossible, to establish discrete racial categories in humans. Races, as we have seen, are geographically defined aggregates of local populations. Human populations are no longer sharply separated geographically from one another. Multiple migrations of peoples and innumerable intermarriages have tended to blur the genetical contrasts between populations. The boundaries of human races, if they can be delimited at all, are at best fuzzy, ever-shifting with time.

The term *race* is regrettably one of the most abused words in the English vocabulary. The biologist views a race as synonymous with a geographical subspecies; a race or subspecies is a genetically distinguishable subgrouping of a species. It is exceedingly important to recognize that a race is *not* a community based on language, literature, religion, nationality, or customs. There are Aryan languages, but there is no Aryan race. Aryans are peoples of diverse genetical makeups who speak a common tongue (Indo-European). *Aryan* is therefore nothing more than a linguistic designation. In like manner, there is a Jewish religion, but not a Jewish race. And there is an Italian nation, but not an Italian race. A race is a reproductive community of individuals occupying a definite region, and in one and the same geographical region may be found Aryans, Jews, and Italians. Every human population today consists of a multitude of diverse genotypes. A "pure" population or race, in which all members are genetically alike, is a myth and an absurdity.

Selected Readings

Ayala, F. J. 1975. Genetic differentiation during the speciation process. *Evolutionary Biology* 8:1–78.

———. 1978. The mechanisms of evolution. *Scientific American,* September, pp. 56–69.

Dobzhansky, T. 1951. *Genetics and the origin of species.* New York: Columbia University Press.

Goldsby, R. A. 1971. *Race and races.* New York: Macmillan Co.

King, J. C. 1971. *The biology of race.* New York: Harcourt Brace Jovanovich.

Mayr, E. 1942. *Systematics and the origin of species.* New York: Columbia University Press.

Mead, M.; Dobzhansky, T.; Tobach, E.; and Light, R. E., eds. 1968. *Science and the concept of race.* New York: Columbia University Press.

Montagu, A. 1964. *The concept of race.* New York: Free Press of Glencoe.

———. 1965. *Man's most dangerous myth: the fallacy of race.* New York: Macmillan Co.

Simpson, G. G. 1953. *The major features of evolution.* New York: Columbia University Press.

Solbrig, O. T., and Solbrig, D. J. 1979. *Introduction to population biology and evolution.* Reading, Mass.: Addison-Wesley.

Stebbins, G. L. 1966. *Processes of organic evolution.* Englewood Cliffs, N.J.: Prentice-Hall.

———. 1971. *Chromosomal evolution in higher plants.* London: Edward Arnold.

Wallace, B., and Srb, A. 1964. *Adaptation.* Englewood Cliffs, N.J.: Prentice-Hall.

10
Cataclysmic Evolution

The process leading to the formation of a new species generally extends over a great reach of time. As we have seen, the origin of new species involves a long period of geographical isolation and the long-term influences of natural selection. A sudden and rapid emergence of a new kind of organism is scarcely imaginable. Yet, a natural mechanism does exist whereby a new species can arise rather abruptly. The process is associated with the phenomenon of *polyploidy*, or the multiplication of the chromosome complement of an organism. Species formation through polyploidy has occurred almost entirely, if not exclusively, in the plant kingdom. Many of our valuable cultivated crop plants, such as wheat, oats, cotton, tobacco, and sugar cane, trace their origin to this cataclysmic or explosive type of evolution.

Wheat

The domestic wheats and their wild relatives have an intriguing evolutionary history. There are numerous species of wheat, all of which fall into three major categories on the basis of their chromosome numbers. The most ancient type is the small-grain einkorn wheat, containing 14 chromosomes in its body (somatic) cells. There are two species, one wild and the other cultivated, both of which may be found growing in the hilly regions of southeastern Europe and southwestern Asia. Cultivated einkorn has slightly larger kernels than the wild form, but the yields of each are low and the grain is used principally for feeding cattle and horses.

Another assemblage of wheat, once widely grown, is the emmer series, of which there are at least six species. The chromosome number in the nuclei of somatic cells of emmer wheats is 28. These varieties, found in Europe and the United States, are used today principally as stock feed, although one of them, called durum wheat, is of commercial value in the production of macaroni and spaghetti.

Table 10.1
Species of Wheat (Triticum)

14 Chromosomes	28 Chromosomes	42 Chromosomes
T. aegilopoides (Wild Einkorn)	*T. dicoccoides* (Wild Emmer)	*T. aestivum* (Bread Wheat)
T. monococcum (Cultivated Einkorn)	*T. dicoccum* (Cultivated Emmer)	*T. sphaerococcum* (Shot Wheat)
	T. durum (Macaroni Wheat)	*T. compactum* (Club Wheat)
	T. persicum (Persian Wheat)	*T. spelta* (Spelt)
	T. turgidum (Rivet Wheat)	*T. macha* (Macha Wheat)
	T. polonicum (Polish Wheat)	

The most recently evolved, and by far the most valuable agriculturally, are the bread wheats. The bread wheats have not been known to occur in the wild state; all are cultivated types. The bread wheats have 42 chromosomes. These wheats, high in protein content, comprise almost 90 percent of all the wheat harvested in the world today.

The various species of wheats thus fall into three major groups, with 14, 28, and 42 chromosomes, respectively. A list of the representatives of these three groups is given in table 10.1.

Origin of Wheat Species

Virtually all authorities are agreed on the sequence of evolutionary events depicted in figure 10.1. Einkorn wheat, possessing a chromosome number of 14, was doubtless one of the ancestral parents of the 28-chromosome emmer assemblage. A most remarkable, but generally accepted, thesis is that the other parent was not a wheat at all, but rather *Aegilops speltoides*, a wild grass with 14 chromosomes. This wild grass parent occurs as a common weed in the wheat fields of southwestern Asia. The cross of einkorn wheat and the wild grass would yield an F_1 hybrid that possesses 14 chromosomes, 7 from each parent. We may designate the 7 chromosomes from one parent species as set (or genome) A, and the 7 from the other parent species as set (or genome) B. Accordingly, the F_1 hybrids would have the AB genomes.

If the chromosome complement in the hybrid accidentally doubled, then the hybrid would contain 28 instead of 14 chromosomes and pass on the doubled set of chromosomes to its offspring. Such an event, strange as it may seem, accounts for the emergence of the 28-chromosome emmer wheat. This new species is characterized as having the $AABB$ genomes.

Figure 10.1   Evolution of wheat. Emmer wheat resulted from the hybridization of einkorn wheat with a wild grass, *Aegilops speltoides*. The common bread wheat is the product of hybridization of emmer wheat with a useless and noxious species of wild grass, *Aegilops squarrosa*. The hybridizations were followed by chromosome doubling, a phenomenon discussed in the text.

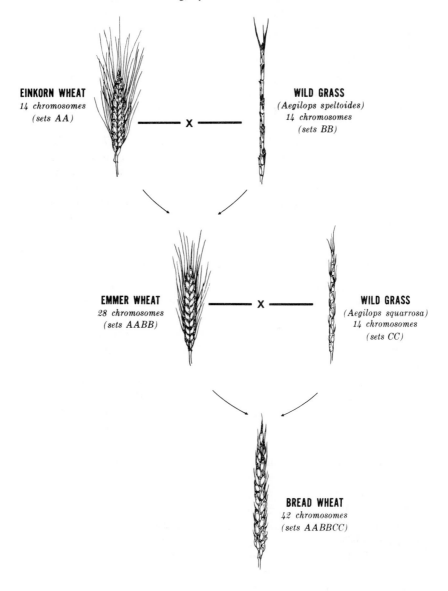

EINKORN WHEAT
14 chromosomes
(sets AA)

X

WILD GRASS
(Aegilops speltoides)
14 chromosomes
(sets BB)

EMMER WHEAT
28 chromosomes
(sets AABB)

X

WILD GRASS
(Aegilops squarrosa)
14 chromosomes
(sets CC)

BREAD WHEAT
42 chromosomes
(sets AABBCC)

In turn, the 28-chromosome emmer wheat was the ancestor of the 42-chromosome bread wheat. In the early 1900s, the British botanist John Percival hazarded the opinion that the bread wheat group arose by hybridization of a species of wheat of the emmer group (28 chromosomes) and goat grass, *Aegilops squarrosa*, a useless weed commonly found growing in wheat fields in the Mediterranean area. Although this startling suggestion was initially viewed with skepticism, it is currently conceded that Percival was correct. *Aegilops squarrosa* possesses 14 chromosomes, and thus would transmit 7 of its chromosomes (set *C*) to the hybrid. The hybrid would contain 21 chromosomes (sets *ABC*), having received 14 (sets *AB*) from its emmer wheat parent. The subsequent duplication in the hybrid of each chromosome set provided by the parents would result in a 42-chromosome wheat species (*AABBCC*).

The initial F_1 hybrid between einkorn wheat and *Aegilops speltoides* (or between emmer wheat and *Aegilops squarrosa*) is sterile, but when the chromosome complement doubles, then a fully fertile species arises. Is it to be expected that the F_1 hybrid would be sterile? And what would account for the fertility of the hybrid when chromosome doubling occurred? This requires a deeper look into the phenomenon of polyploidy, to which we shall now turn.

Mechanism of Speciation by Polyploidy

Figure 10.2 illustrates the underlying basis of the fertility of a formerly sterile hybrid resulting from the doubling of its chromosome number. For ease of presentation, the parental species are shown with a small number of chromosomes, 6 and 4, respectively. It should be noted that the chromosomes are present in pairs. The members of each pair are alike or homologous, but each pair is distinguishable from the other. Thus, the 6-chromosome parent possesses three different pairs of chromosomes; the 4-chromosome parent, two different pairs.

The gametes, egg cells and pollen cells (sperm), are derived by cell divisions of a special kind, called meiosis. One of the essential features of the process of meiosis is that the members of each pair are attracted to each other and come to lie side by side in the nucleus. The meiotic cell divisions are intricate, but the pertinent outcome is the separation of like, or homologous, chromosomes, such that each gamete comes to possess only one member of each pair of chromosomes. Each gamete is said to contain a *haploid* complement of chromosomes, or half the number found in a somatic cell. The latter, in turn, is described as being *diploid* in chromosome number.

The sterility of the first-generation hybrid is now comprehensible. There are simply no homologous chromosomes in the F_1 hybrid. Each chromosome lacks a homologue to act as its pairing partner at meiosis. The process of meiosis in the hybrid is chaotic; the chromosomes move at random into the gametes. The eggs and pollen cells typically contain an odd assortment of chromosomes and are nonfunctional.

Figure 10.2 **Sequence of events** leading to a new, fertile species from two old species by hybridization and polyploidy. The F_1 hybrid plant derived from a cross of the two parental species is sterile. The F_1 hybrid may occasionally produce viable gametes when all chromosomes fortuitously enter a gamete during the process of meiosis. The fusion of such gametes leads to a new form of plant, which contains two complete sets of chromosomes (one full set of each of the original parents.)

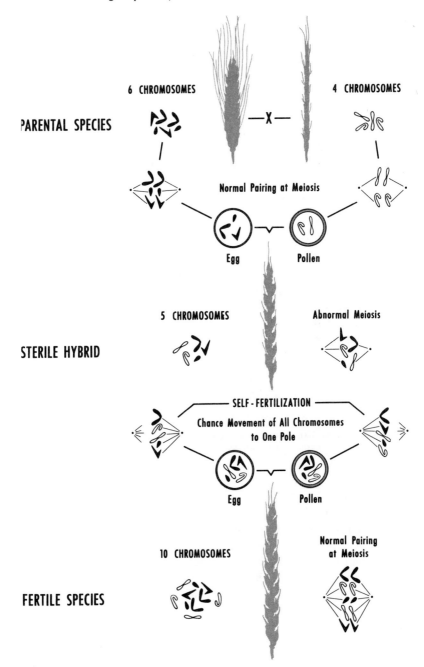

Occasionally, by sheer chance, a few gametes might be produced by the F_1 hybrid that contain all the chromosomes (fig. 10.2). These gametes would be functional, and the fusion of such sex cells would give rise to a plant that contains twice the number of chromosomes that the first-generation hybrid possessed. The plant actually would contain two complete sets of chromosomes; that is, the full diploid complement of chromosomes of each original parent. Such a double diploid is termed a polyploid—more specifically, a *tetraploid*.

The tetraploid hybrid would resemble the first-generation hybrid, but the plant as a whole would be larger and somewhat more robust as a consequence of the increased number of chromosomes. More importantly, the tetraploid hybrid would be fully fertile. The meiotic divisions would be normal, since each chromosome now has a regular pairing partner during meiosis (fig. 10.2). The tetraploid hybrid is a true breeding type; it can perpetuate itself indefinitely. It is, however, reproductively incompatible with its original parental species. If it were to cross with its original parental species, the offspring would be sterile. Hence, the tetraploid hybrid is truly a new distinct species. In but a few generations, we have witnessed essentially the fusion of two old species to form a single derived species.

Experimental Verification

To return to our wheat story, bread wheat (fig. 10.1) contains the chromosome sets of three diploid species—the AA of einkorn wheat, the BB of *Aegilops speltoides*, and the CC of *Aegilops squarrosa*. (The third set is typically referred to by botanists as the D genome. It is an accident of nomenclature that the third genome received the letter D, rather than C.) Technically, then, the bread wheat contains six sets of chromosomes; it is a *hexaploid*. The relationships of the three major groups of wheats can be shown as follows:

Einkorn $= 14 = AA =$ diploid
Emmer $= 28 = AABB =$ tetraploid
Bread $= 42 = AABBCC =$ hexaploid

Experimental proof was lacking at the time John Percival proposed that the bread wheats originated from hybridization between the emmer wheat and goat grass, followed by chromosome doubling in the hybrid. Verification awaited an effective method of artificially inducing diploid cells to become polyploid. The search for an efficient chemical inducing agent culminated in the discovery in the late 1930s of colchicine, a substance obtained from the roots of the autumn crocus plant. Treatments of diploid plant cells with colchicine result in a high percentage of polyploid nuclei in the treated plant cells. Colchicine acts on the spindle apparatus of a dividing cell, and prevents a cell from dividing into two daughter halves. The treated undivided cell contains two sets of daughter chromosomes, which ordinarily would have separated from each other had cell division not been impeded. The cell thus comes to possess twice the usual number of chromosomes.

The experimental production of polyploid cells through the application of colchicine paved the way for studies on wheat by E. S. McFadden and E. Sears of the United States Department of Agriculture. These investigators successfully hybridized a tetraploid species of emmer wheat with the diploid wild grass *Aegilops squarrosa*. The chromosome number in the hybrid was doubled by treatment with colchicine. The synthetic hexaploid hybrids were similar in characteristics to natural hexaploid species of bread wheat, and produced functional gametes. At almost the same time, Hitoshi Kihara of Japan obtained a comparable hexaploid wheat species, which spontaneously and naturally had become converted from a sterile hybrid to a fertile hybrid. Kihara's work reinforced the notion that doubling of chromosomes can occur accidentally.

To complete the proof, McFadden and Sears crossed their artificially synthesized hexaploid wheat species with one of the naturally occurring bread wheat *Triticum spelta*. Fully fertile hybrids resulted, removing any doubt that wild grass, a noxious weed, is indeed a parental ancestor of the bread wheats.

Cotton

The story of the cataclysmic evolution of cotton (*Gossypium*) has been partially unraveled by botanical investigators. The evolution of cotton poses some interesting, but unresolved, problems.

Cotton is widely distributed throughout the world, and occurs in both the wild and the cultivated state. The cultivated types in the Americas are represented by *Gossypium barbadense*, the prominent cotton of South America, and *Gossypium hirsutum*, grown mainly in Central America and the United States. These two cultivated species of American cotton are of particular interest in that each possess 52 chromosomes (or 26 pairs) in their cells. Thirteen of these pairs of chromosomes are small and resemble those in wild diploid cotton species still found growing in the Americas; the other 13 pairs are large and like those of diploid cotton species native to the Old World. The Old World cotton ancestor contains the A chromosome set (or genome); the wild American ancestor possesses the D genome. Unquestionably, the present-day American cultivated cotton resulted from a cross between the Old World cotton and the wild American cotton ($AA \times DD$), with subsequent natural duplication of the chromosome complement in the hybrid (AD to $AADD$). The cultivated American cottons are thus tetraploid progenies of two diploid species. The question as to when and where the two diploid progenitors met and hybridized is a thorny one.

Available evidence indicates that the American tetraploid species arose by hybridization in the coastal valleys of Peru in western South America. Many botanists contend that the Old World diploid parent came from southern Asia, having reached South America by dispersal across the Pacific Ocean. The seeds of cotton, however, are not adapted to transportation over great distances by either water or wind. One authority on cotton, S. C. Harland, has suggested that the Old World parent crossed the Pacific by a land bridge in late Cretaceous times, about 100 million years ago. Most modern geologists dismiss as unreasonable the once popular notion of an ancient Pacific land bridge.

Other botanists, particularly J. B. Hutchinson, R. A. Silow, and S. G. Stephens, have argued that the Asiatic diploid cotton was introduced within historic times by civilized man. Early nomadic man from Asia carried the seeds of his crop plants across the Pacific to South America. From hybridizations of the transplanted Old World cotton with wild cotton of the valleys of Peru emerged the superior tetraploid plant. This tetraploid cotton was subsequently introduced into Central America, and then spread to the United States. The South American tetraploids are no longer reproductively compatible with their northern counterparts; hence, the presence today, as remarked earlier, of two distinct species of cultivated cotton in the Americas, *Gossypium barbadense* and *Gossypium hirsutum*.

The reconstruction of the past history of the cultivated American cotton is far from complete. There exists a primitive diploid species of cotton in south Africa, regarded as the forebear of the Old World diploid cottons. The Asiatic cottons were probably derived from the African type. Recently, there has been speculation that the African species of cotton, rather than the Asiatic species, was the immediate parent of the American cotton. This would be in accord with Wegener's theory of continental drift. In 1912, the Austrian meteorologist and Arctic explorer Alfred Wegener proposed that the earth's continents had once been a huge land mass (a supercontinent) and have reached their present geographical positions by splitting up and drifting across the ocean floors (fig. 10.3). Initially, Wegener's theory was ridiculed. At a meeting of the American Philosophical Society in Philadelphia in the 1920s, the key speaker pronounced Wegener's thesis "utter, damned rot." Today, the idea of continental drift has been revived and has gained a high measure of scientific respectability. Indeed, we now recognize that Wegener was essentially correct.

Cotton and Continental Drift

The original single land mass, called *Pangaea*, split and began to move apart about 180 to 200 million years ago. The supercontinent first split into two blocks: *Laurasia* in the Northern Hemisphere, consisting of North America, Europe, and Asia; and *Gondwanaland* in the Southern Hemisphere, made up of Africa, South America, Antarctica, Australia, and India (which then lay far south of Asia). South America began sliding westward about 135 million years ago when a rift appeared between South America and Africa. Antarctica parted from Africa, and India broke free to move northerly toward Asia. North America separated from Europe only about 80 million years ago. It is of interest that India collided with Asia less than 40 million years ago. The collision uplifted the high Tibetan Plateau and raised the Himalayas.

Figure 10.3   Wegener's theory of continental drift. The present continents were joined in a single land mass (*Pangaea*) before the start of the Mesozoic era (about 200 million years ago). The hybridization of two African diploid species of cotton to give rise to the American tetraploid species of cotton may have occurred before the continents drifted apart.

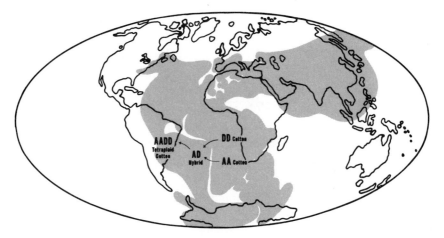

The concept of continental drift serves to explain the present similarity in certain animals and plants in regions of Africa and South America. With respect to cotton, consideration may be given to the possibility that an ancestral diploid cotton species of south African (genome A) at one time overlapped the distribution of another ancestral species (genome D) in that part of western Africa adjacent to the American continent before continental drift (fig. 10.3). In the area of contact, hybridization between the two diploid species may have occurred naturally, giving rise to the American tetraploid species that became isolated in northern South America at the time of continental separation. This hypothesis, although provocative, limps a little. Africa and South America began to separate 135 million years ago, and perhaps earlier. The fossil history of flowering plants does *not* extend into the geologic past farther back than the Cretaceous period, about 135 million years ago (see chapter 14). It seems unlikely that any of the cotton species could have been in existence prior to the separation of the continents when there is essentially no fossil record of flowering plants before the Cretaceous. The hypothesis could be salvaged by assuming that there was a long pre-Cretaceous evolution of flowering plants, but that pre-Cretaceous conditions were not favorable for the preservation of their fossils.

Selected Readings

Briggs, D., and Walters, S. M. 1969. *Plant variation and evolution.* New York: McGraw-Hill Book Co.

Curtis, B. C., and Johnston, D. R. 1969. Hybrid wheat. *Scientific American*, May, pp. 21–29.

Grant, V. 1971. *Plant speciation.* New York: Columbia University Press.

Hurley, P. M. 1968. The confirmation of continental drift. *Scientific American*, April, pp. 53–64.

Kurten, B. 1969. Continental drift and evolution. *Scientific American*, March, pp. 54–64.

Mangelsdorf, P. C. 1953. Wheat. *Scientific American*, July, pp. 50–59.

Solbrig, O. T. 1970. *Principles and methods of plant biosystematics.* New York: Macmillan Co.

Stebbins, G. L. 1950. *Variation and evolution in plants.* New York: Columbia University Press.

————. 1951. Cataclysmic evolution. *Scientific American*, April, pp. 54–59.

11
Adaptive Radiation

The capacity of a population of organisms to increase its numbers is largely governed by the availability of resources—food, shelter, and space. The available supply of resources in a given environment is limited, whereas the organism's innate ability to multiply is unlimited. A particular environment will soon prove to be inadequate for the number of individuals present. It might thus be expected that some individuals would explore new environments where competition for resources is low. The tendency of individuals to exploit new opportunities is a factor of major significance in the emergence of several new species from an ancestral stock. The successful colonization of previously unoccupied habitats can lead to a rich multitude of diverse species, each better fitted to survive and reproduce under the new conditions than in the ancestral habitat. The spreading of populations into different environments accompanied by divergent adaptive changes of the emigrant populations is called *adaptive radiation*.

Galápagos Islands

One of the biologically strangest, yet fascinating, areas of the world is an isolated cluster of islands of volcanic origin in the eastern Pacific, the Galápagos Islands. These islands, which Darwin visited for five weeks in 1835, lie on the equator 600 miles west of Ecuador (fig. 11.1). The islands are composed wholly of volcanic rock; they were never connected with the mainland of South America. The rugged shoreline cliffs are of gray lava and the coastal lowlands are parched, covered with cacti and thorn brushes. In the humid uplands, tall trees flourish in rich black soil.

Giant land-dwelling tortoises still inhabit these islands. After many years of being needlessly slain by pirates and whalers, these remarkable animals now live protected in a sanctuary created in 1959. Still prevalent on the islands are the world's only marine iguana and its inland variety, the land iguana

Figure 11.1   Galápagos Islands ("Enchanted Isles") in the Pacific Ocean, 600 miles west of Ecuador. Darwin had explored this cluster of isolated islands, and found a strange animal life, a "little world within itself." The four stamps shown were issued by Ecuador to commemorate the centenary of Darwin's visit in 1835.

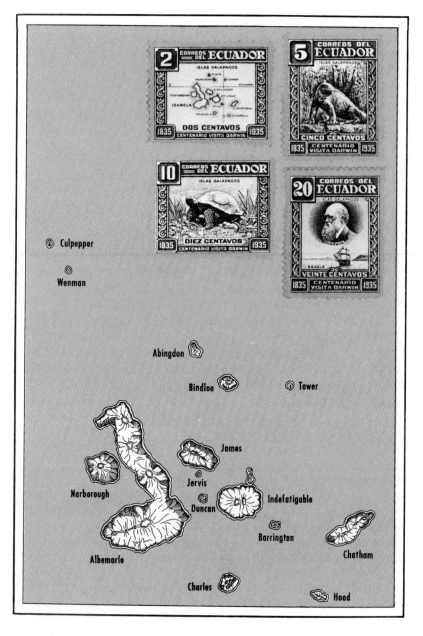

Figure 11.2 **Land iguana** on one of the Galápagos Islands. Despite their horrendous appearance, these bizarre inland lizards are mild, torpid, and vegetarians. They feed on leaves and cactus plants. (American Museum of Natural History.)

(fig. 11.2). These two species of prehistoric-looking lizards are ancient arrivals from the mainland. The marine forms occur in colonies on the lava shores, and swim offshore to feed on seaweed. The land iguana lives on leaves and cactus plants. Cactus fills most of the water needs of the land iguana.

At least 85 different kinds of birds have been recorded on the islands. These include rare cormorants that cannot fly, found only on Narborough Island, and flamingos, which breed on James Island. Of particular interest are the small dark finches. These dusky birds exhibit remarkable variations in the structure of the beak and in feeding habits. The finches afford an outstanding example of adaptive radiation. It was the marked diversity within this small group of birds that gave impetus to Darwin's evolutionary views. Darwin had correctly surmised that the diverse finches were modified descendants of the early, rather homogeneous, colonists of the Galápagos. These birds are now appropriately called "Darwin's finches."

Darwin's Finches

The present-day assemblage of Darwin's finches descended from small sparrowlike birds that once inhabited the mainland of South America. The ancestors of Darwin's finches were early migrants to the Galápagos Islands, and

Figure 11.3   Representatives of Darwin's finches. There are 14 species of Darwin's finches, confined to the Galápagos with the exception of one species that inhabits Cocos Island. Closest to the ancestral stock are the six species of ground finches, primarily seed-eaters. The others evolved into eight species of tree finches, the majority of which feed on insects.

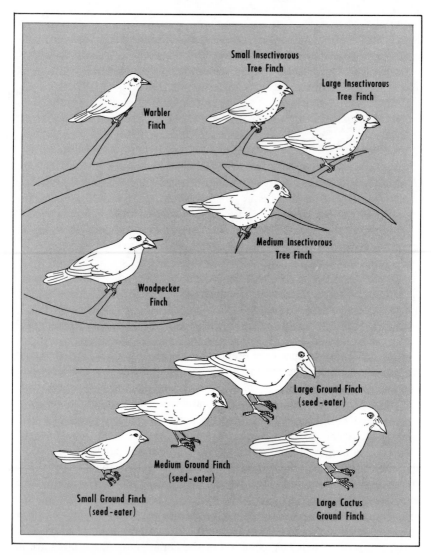

probably the first land birds to reach the islands. These early colonists have given rise to 14 distinct species, each well adapted to a specific niche (fig. 11.3). Thirteen of these species occur in the Galápagos; one is found in the small isolated Cocos Island, northeast of the Galápagos. Not all 13 species are found on each island.

The most striking differences among the species are in the sizes and shapes of the beak, which are correlated with marked differences in feeding habits.

Six of the species are ground finches, with heavy beaks specialized for crushing seeds. Some of the ground finches live mainly on a diet of seeds found on the ground; others feed primarily on the flowers of prickly pear cacti. The cactus eaters possess decurved, flower-probing beaks. Their beaks are thicker than those of typical flower-eating birds.

All the other species are tree finches, the majority of which feed on insects in the moist forests. One of the most remarkable of these tree dwellers is the woodpecker finch. It possesses a stout, straight beak, but lacks the long tongue characteristic of the true woodpecker. Like a woodpecker, it bores into wood in search of insect larvae, but then it uses a cactus spine or twig to probe out its insect prey from the excavated crevice. Equally extraordinary is the warbler finch, which resembles in form and habit the true warbler. Its slender, warblerlike beak is adapted for picking small insects off bushes. Occasionally, like a warbler, it can capture an insect in flight.

Factors in Diversification

No such great diversity of finches can be found on the South American mainland. In the absence of vacant habitats on the continent, the occasion was lacking for the mainland birds to exploit new situations. However, given the unoccupied habitats on the Galápagos Islands, this was opportunity for the invading birds to evolve in new directions. In the absence of competition, the colonists occupied several ecological habitats, the dry lowlands as well as the humid uplands. The finches adopted modes of life that ordinarily would not have been opened to them. If true warblers and true woodpeckers had already occupied the islands, it is doubtful that the finches could have evolved into warblerlike and woodpeckerlike forms. Thus, *a prime factor promoting adaptive radiation is the absence of competition.*

The emigration of the ancestral finches from the mainland was assuredly not conscious or self-directed. The dispersal of birds from the original home was at random, resulting from the pressure of increasing numbers on the means of subsistence. By chance, some of the birds reached the Galápagos Islands. The original flock of birds that fortuitously arrived at the islands was but a small sample of the parental population, containing at best a limited portion of the parental gene pool. It may be that only a small amount of genetic variation was initially available for selection to work on. What evolutionary changes occurred at the outset were mainly due to random survival (genetic drift). However, the chance element would become less important as the population increased in size. Selection unquestionably became the main evolutionary agent, molding the individual populations into new shapes by the preservation of new favorable mutant or recombination types. More than one island was colonized, and the complete separation of the islands from each other promoted the genetic differentiation of each new local population.

The Galápagos Islands are the tips of enormous volcanoes, most of which rise from 7,000 to 10,000 feet above the floor of the sea. It is of especial interest that five distinct varieties of tortoises occupy regions corresponding to the five principal volcanic tops of the large Island of Albemarle (fig. 11.4).

Figure 11.4   Distribution of the five subspecies of the giant land tortoise (*Geochelone elephantopous*) on Albemarle Island. Each of the five subspecies inhabits one of the five major volcanoes on Albemarle. Each subspecies evolved independently on the separate volcanic tops of the island.

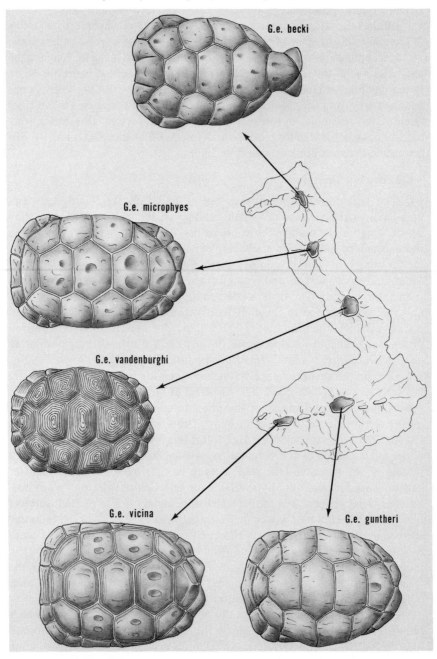

The five different forms of tortoises evolved independently of each other at a time when Albemarle Island was so deeply submerged that only its five major craters were above water. The subsequent lowering of the sea level in the geologic past had the effect of uniting the five volcanoes into a single land mass. Even to this day the five volcanoes are separated from each other by deep valleys. There is no intermingling of the five varieties of tortoises. This intriguing geologic circumstance serves to document the importance of long-term isolation for the genetic differentiation of the separated populations.

Competitive Exclusion

When two populations of different species are obliged (under experimental conditions) to use a common nutrient, the two will compete with each other for the common resource. Under competition for the identical needs, only one of the two species populations will survive; the other will be eliminated or excluded by competition. This is the principle of *competitive exclusion.* It is also known as *Gause's principle,* after the Russian biologist G. F. Gause, who first demonstrated experimentally that under controlled laboratory conditions, one of the two competing species perishes.

In 1934, Gause studied the interactions under carefully controlled culture conditions of two protozoan species, *Paramecium caudatum* and *Paramecium aurelia.* When each species was grown separately in a standard medium in a test tube containing a fixed amount of bacterial food, each species flourished independently. When the two species were placed together in the same culture vessel, however, the growth of *P. caudatum* gradually diminished until the population became eliminated. In enforced competition for the same limited food supply, *P. aurelia* was the more successful species.

Gause enunciated his principle primarily on the basis of observations of "bottle populations" in the laboratory. The competition experiments reveal that two species populations cannot exist together if they are competing for precisely the same limited resource. Alternatively, if two species in nature were to occupy the same habitat, the expectation is that each would have different ecological requirements, even though the degree of difference is slight. Ecologists have demonstrated the validity of this view. In virtually every natural situation carefully examined, two co-inhabiting species have been found to differ in some requirement. The heterogeneous resources of the environment in a given locality are typically partitioned among the co-inhabiting species to minimize direct competition and enable the two (or more) species to coexist.

Most of the Galápagos Islands are occupied by more than one species of finch. On islands where several species of finches exist together, we find that each species is adapted to a different ecological niche. The three common species of ground finches—small (*Geospiza fuliginosa*), medium (*Geospiza fortis*), and large (*Geospiza magnirostris*)—occur together in the coastal lowlands of several islands. Each species, however, is specialized in feeding on a seed of a certain size. The small-beaked *Geospiza fuliginosa*, for example, feeds on small grass seeds, whereas the large-beaked *Geospiza magnirostris* eats large, hard fruits. Different species, with different food requirements, can thus exist together in an environment with varied food resources.

Coexistence

Gause's principle may be stated in the following form: No two species with identical requirements can continue to exist together. However, it is exceedingly unlikely that two species in nature would have *exactly* the same requirements for food and habitat. The sum total of environmental requirements for a species to thrive and reproduce has been termed the *niche* of that species population. The term *niche*, as ecologists use it, is more than simply the physical space that the species population occupies. It is essentially the way of life peculiar to a given species: its structural adaptations, physiological responses, and behavior within its habitat. Experience has shown that the likelihood of two species having identical niches is almost nil.

Direct evidence of the process of competition between species in nature is difficult to obtain. Observable competitive interaction is a relatively fleeting stage in the relation of two species populations. What the ecologist observes is the end result of competitive contact, when the actual or potential competitors have become differentially specialized to exploit different components of a local environment and accordingly live side by side. The outcome, then, of incipient competitive interaction is the *avoidance* of competition through differential specialization, or in the terminology of ecologists, through *niche diversification*.

Indirect arguments have been used to support the view that two closely related species populations come to exploit different ecological niches in the same locality after a beginning of competitive interaction. We may envision a situation in which two closely related species, with almost similar ecological requirements, expand their geographic ranges and meet in a common habitat. It may be presumed that each of the two species populations has appreciable genetic variability, and that the resources in the common habitat are varied. The two species populations would initially compete for suitable ecological niches in the new common habitat. However, we can expect that the members of the two species will ultimately become so different in structure and behavior that each species will become specialized to use different components of the environment. In other words, if genetic differences in morphological and behavioral characteristics tend at first to reduce competition between the two species, then subsequently natural selection will act to augment the differences between the two competing species. It is especially noteworthy that the differences between the two species become more pronounced as a consequence of selection *reducing* competition rather than *intensifying* competition.

A striking case of niche diversification has been described by David Lack for certain species of *Geospiza* in the Galápagos. Where two species occur together on an island, there is a conspicuous difference in the size of the beaks and food habits. Where either species exists alone on other islands, the beak is adapted to exploit more than one food resource. Thus, on the island of Tower (fig. 11.5), the cactus-feeding *Geospiza conirostris* and the large seed-

Figure 11.5 Niche diversification. In the absence of competition, the ground finch *Geospiza conirostris* on Hood Island has evolved a large beak that enables it to exploit a variety of food items (seed and cacti). On the island of Tower, the same ground finch *Geospiza conirostris* has evolved a specialized beak adapted solely for cactus feeding, thus reducing or avoiding competition for food resources with the seed-eating ground finch *Geospiza magnirostris*. The latter species does not exist on Hood Island.

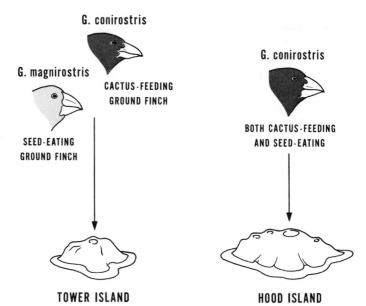

eating ground finch *Geospiza magnirostris* live side by side. The former species occurs also on the island of Hood, but the latter species is absent, presumably having failed to invade or reach the island. In the absence of competition, *Geospiza conirostris* on Hood Island has evolved a larger beak that is adapted to feeding on both cactus and seeds. The sharp separation in beak size of the two species on Tower Island is understandable if we assume that competition initially fostered the differentiation of the beaks to permit each of the two species to adapt to a limited or restricted range of the available food resources. Each species is now genetically specialized in food habits, and competition between the two is now avoided.

There are many examples to illustrate how two or more species avoid competition. Among warblers that inhabit the spruce forests of Maine, each species confines its feeding to a particular region of the spruce tree. The myrtle warbler, for example, preys on insects in and below the spruce tree, whereas the Blackburnian warbler prefers those insects on the exterior leaves of the top of the tree. They can exist together because they use different resources of the same tree.

We may conclude by stating, paradoxically, that competition between two species populations achieves the avoidance or reduction of further competition, and not an intensification. In natural populations, coexistence of two species, rather than competitive exclusion, is the general rule.

Geographic Speciation

In an earlier chapter we said that geographic or spatial isolation is an important, if not essential, condition for the formation of a new animal species. When two populations are separated by a geographical barrier, each may diverge independently of the other. The distribution of insectivorous tree finches on some of the Galápagos Islands attests to the significance of geographical isolation in the speciation process.

One particular assemblage of tree finches has a widespread distribution (fig. 11.6). These finches now show a discontinuous pattern of geographical variation. If we were able to gaze into the evolutionary past, we would probably witness the series of events depicted in figure 11.6. At one time, we would be able to recognize three well-defined geographical groups, which, for simplicity of presentation, may be designated *A, B,* and *C.* We might then subsequently observe the Albemarle form (*C*) spreading to Charles Island and establishing a colony, which, in time, differentiated into group *C'.* The Albemarle finch today does show a close resemblance to the Charles form, which suggests that the latter was derived from the former. If we were to stop the evolutionary clock at this time, we would properly regard this assemblage of finches as a single species, subdivided into four well-marked geographic races. In fact, this assemblage has been designated as one species, *Camarhynchus psittacula.*

At the present time, we find that Charles Island is inhabited not only by group *C'* but also by a newly established colony, whose members are indistinguishable from those of group *A.* Undoubtedly, the *A* group on Charles Island comprises recent arrivals from one of the central islands, Indefatigable or Barrington.

The more recent immigrants, group *A,* on Charles occur together with the older colonists, group *C',* but they do *not* interbreed. Thus, the two groups, having arisen initially from separate islands, had developed sufficient genetic differences in isolation to remain distinct when they met on Charles Island. Groups *A* and *C'* are two separate species, actually designated as *Camarhynchus psittacula* and *Camarhynchus pauper,* respectively.

This situation raises some interesting questions. If group *C* on Albermarle were now to spread to Abingdon, where *B* occurs, and an exchange of genes did not take place between them, then *B* and *C* would be considered as distinct species. But, as far as we know, *C* has not invaded the territory of *B,* and the proper designations of the two groups cannot be made. Most evolutionists would probably regard *B* and *C* as races, or subspecies, of the same species, *Camarhynchus psittacula.* Regardless of the correct terminology, the important consideration is the demonstration that species formation is preceded by geographical separation.

Classification

The original ancestral stock of finches on the Galápagos diverged along several different paths. The pattern of divergence is reflected in the biologists' scheme of classification of organisms. All the finches are related to one another, but

Figure 11.6 Historical events leading to the existence today of two distinct species of tree finches, *Camarhynchus psittacula* (designated *A*) and *Camarhynchus pauper (C')*, which occur together on Charles Island. Charles Island had been colonized by finches on two separate occasions. The earlier immigrant came from Albemarle Island in the distant past; the later immigrant was a more recent arrival from Indefatigable or Barrington. The two immigrating groups, although initially only races of the same species, had accumulated sufficient genetic differences in isolation that they did not interbreed when they met on Charles Island. Thus, they are separate species. (Based on extensive studies by David Lack.)

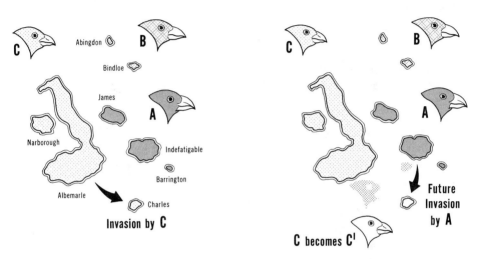

DISTANT PAST

Abingdon

Bindloe

James

Narborough

Indefatigable

Barrington

Albemarle

Charles

Invasion by **C**

RECENT PAST

C becomes **C'**

Future Invasion by **A**

PRESENT

Two Distinct Species

Figure 11.7 Evolutionary tree of Darwin's finches, graphically expressing what is known or surmised as to the degree of relationship or kinship between the different species of finches. Darwin's finches evolved from a common stock, and are represented today by 14 species. The species are assembled in four genera, of which two (Certhidea and Pinaroloxias) contain only a single species each. The species associated together in a genus have significant common attributes, judged to denote evolutionary affinities. (Based on the findings of David Lack.)

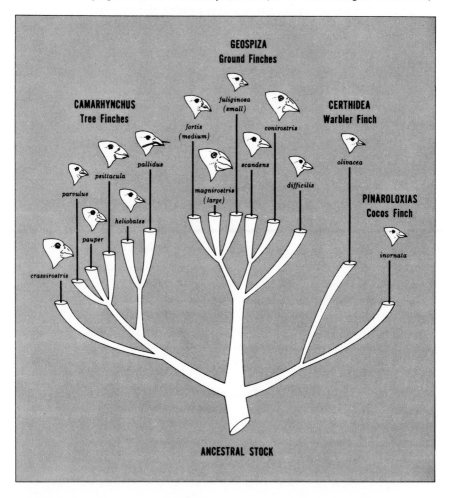

the various species of ground finches obviously are more related by descent to one another than to the members of the tree-finch assemblage. As a measure of evolutionary affinities, the ground finches are grouped together in one genus (*Geospiza*) and the tree finches are clustered in another genus (*Camarhynchus*). The different lineages of finches are portrayed in figure 11.7. It should be clear that our classification scheme is nothing more than an expression of evolutionary relationships between groups of organisms.

In the next chapter we shall see how adaptive radiation on a much larger scale than that which occurred in the finches led to the origin of radically new assemblages of organisms, distinguishable as orders and classes by the taxonomist.

Selected Readings

Dillon, L. S. 1973. *Evolution: concepts and consequences.* St. Louis: C. V. Mosby Co.

Grant, V. 1963. *The origin of adaptations.* New York: Columbia University Press.

Hamilton, T. H. 1967. *Process and pattern in evolution.* New York: Macmillan Co.

Lack, D. 1947. *Darwin's finches.* London: Cambrige University Press.

————. 1953. Darwin's finches. *Scientific American,* April, pp. 66–72.

Mayr, E. 1963. *Animal species and evolution.* Cambridge, Mass.: Harvard University Press.

Merrell, D. J. 1962. *Evolution and genetics.* New York: Holt, Rinehart & Winston.

Moody, P. A. 1962. *Introduction to evolution.* New York: Harper & Row.

Moore, R., and editors of Time-Life Books. 1968. *Evolution.* New York: Time-Life Books.

Smith, J. M. 1958. *The theory of evolution.* Baltimore: Penguin Books.

12

Major Adaptive Radiations

The diversity of Darwin's finches had its beginning when migrants from the mainland successfully invaded the variety of vacant habitats on the Galápagos Islands. The pattern of adaptive radiation manifested by Darwin's finches has been imitated repeatedly by different forms of life. Organisms throughout the ages have seized new opportunities open to them by the absence of competitors and have diverged in the new environments. The habitats available to Darwin's finches were certainly few in comparison with the enormous range of ecological habitats in the world. The larger the region and the more diverse the environmental conditions, the greater the variety of life.

Approximately 400 million years ago, during a period of history that geologists call the Devonian, the vast areas of land were monotonously barren of animal life. Save for rare creatures like scorpions and millipedes, animal life of those distant years was confined to the water. The seas were crowded with invertebrate animals of varied kinds. The fresh and salt waters contained a highly diversified and abundant assemblage of cartilaginous and bony fishes. The vacant terrestrial regions were not to remain long unoccupied. From one of the many groups of fishes inhabiting the pools and swamps in the Devonian period emerged the first land vertebrate. The initial modest step onto land started the vertebrates on their conquest of all available terrestrial habitats. The origin and diversification of the backboned land dwellers is one of the most striking examples of adaptive radiation on a large scale.

Invasion of Land

Prominent among the numerous Devonian aquatic forms were the lobe-finned fishes, the Crossopterygii, which possessed the ability to gulp air when they rose to the surface. These ancient air-breathing fishes represent the stock from which the first land vertebrates, the amphibians, were derived (fig. 12.1). The

Figure 12.1 **Evolution of land vertebrates in the geologic past.** From air-breathing, lobe-finned fishes (crossopterygians) emerged the first four-footed land inhabitants, the amphibians. Primitive amphibians (labyrinthodonts) gave rise to the reptiles, the first vertebrates to become firmly established on land. The birds and mammals owe their origin to an early reptilian stock (cotylosaurs). An important biological principle reveals itself: each new vertebrate group did not arise from highly developed or advanced members of the ancestral group, but rather from early primitive forms near the base of the ancestral stock. The thickness of the various branches provides a rough measure of the comparative abundance of the five vertebrate groups during geologic history. The Devonian period is often called the *Age of Fishes;* the Mississippian and Pennsylvanian periods (frequently lumped together as the Carboniferous period) are referred to as the *Age of Amphibians;* the Mesozoic era is the grand *Age of Reptiles;* and the Cenozoic era is the *Age of Mammals.*

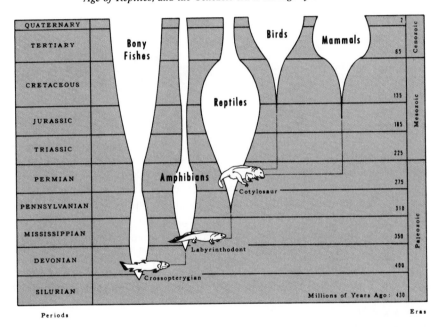

factors that led these ancestral lobe-finned fishes to venture onto land are unknown. The impelling force might have been population pressure or simply the inherent tendency of individuals, particularly of the young, to disperse.

The above hypothesis is admittedly speculative. However, the fact remains that these crossopterygians that emerged on land, though crudely adapted for terrestrial existence, did not encounter any competitors that could immediately spell doom to their awkward initial trial on land. We should note that the lobe-finned fishes did possess certain capacities that would prove to be important under the new conditions of life. Evolutionists speak of such potential adaptive characters as *preadaptations.*

The preadaptations of the lobe-finned fishes included primitive membranous lungs and internal nostrils, both of which are important for atmospheric

Figure 12.2   The evolution of lungs and limbs. The main elements of the limb—humerus (H), radius (R), and ulna (U)—are clearly comparable in a lobe-finned fish and an early amphibian. The irregular outer branches of the fish limb are transformed into a pattern of digits in a land animal. The lungs represent outpocketings from the underside of the pharynx (P) of the digestive tract. The foldings of the internal walls of the lungs in land animals improve the efficiency of the lungs.

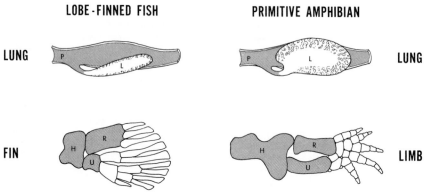

breathing (fig. 12.2). It should be understood that such preadapted characters were not favorably selected with a view to their possible use in some future mode of life. There is no foresight or design in the selection process. A structure does not evolve in advance of impending events. Nor do mutational changes occur in anticipation of some new environmental condition. A trait is selected only when it imparts an advantage to the organism in its immediate environment. Accordingly, lungs in the crossopterygians did not evolve with conscious reference toward a possible future land life, but only because such a structure was important, if not essential, to the survival of these air-breathing fishes in their immediate surroundings. In essence, the selection process does not foresee environments yet to come. The expression "preadaptation" simply signifies that a structure evolves under one set of environmental circumstances and becomes modified later for a totally different function.

The crossopterygians did not, of course, possess typical amphibian limbs. However, their lateral fins contained fleshy lobes, within which were bony elements that were basically comparable to those of a limb of a terrestrial vertebrate. Figure 12.3 shows a restoration of a widespread Devonian form, *Eusthenopteron,* in which the lateral fins had developed into stout, muscular paddles. It bears emphasizing that these primitive amphibians arose from a population of air-breathing fishes that fortuitously had the minimal preadaptive features to survive in the vacant (terrestrial) ecological niche.

Before the close of the Devonian period, the transition from fish to amphibian had been completed. The early land-living amphibians were slim-bodied with fishlike tails, but having limbs capable of locomotion on land. The

Figure 12.3   **Stages in the transition of the lobe-finned fishes into amphibians,** as reconstructed by W. K. Gregory and painted by Francis Lee Jaques. *Bottom,* the primitive Devonian air-breathing crossopterygian, *Eusthenopteron,* floundering on a stream bank with its muscular, paddlelike fins. *Top,* the Pennsylvanian tailed amphibian, *Diplovertebron,* with limbs capable of true locomotion on land. Much of the life of this early tetrapod was spent in the water. (American Museum of Natural History.)

four-footed amphibians flourished in the humid swamps of Mississippian and Pennsylvanian times, but never did become completely adapted for existence on land. All the ancient amphibians, such as *Diplovertebron* (fig. 12.3), spent much of their lives in water, and their modern descendants—the salamanders, newts, frogs, and toads—must return to water to deposit their eggs. Thus, the amphibians were the first vertebrates to colonize land, but were, and still are, only partially adapted for terrestrial life.

Conquest of Land

From the amphibians emerged the reptiles, true terrestrial forms. The appearance of a shell-covered egg, which can be laid on land, freed the reptile from dependence on water. The elimination of a water-dwelling stage was a significant evolutionary advance. The first primitive reptile most likely arose during Carboniferous times, but the fossil beds of this period have yet to reveal the appropriate reptilian ancestor. The ancestral reptile probably possessed the body proportions and well-developed limbs of the more advanced terrestrial forms of amphibians, such as *Seymouria* (fig. 12.4). *Seymouria* is not a reptile, but its skeletal features suggest terrestrial habits. This advanced amphibian may have been a descendant of an earlier amphibian group in the lower Carboniferous that was ancestral to the reptiles.

A key feature of reptiles (and higher vertebrates) is the *amniotic egg* (fig. 12.5). The egg of a reptile (or bird) contains a large amount of nourishing yolk. Moreover, development of the embryo takes place entirely within a thick shell. These circumstances call for a special provision whereby food derived from the yolk and oxygen obtained from an external source can be made accessible to all parts of the developing embryo. The embryo itself constructs a complex system of membranes, known as *extraembryonic membranes,* which serve for protection, nutrition, and respiration. These are the *amnion, chorion, yolk sac, and allantois.*

When fully developed, the amnion is a thin membrane loosely enclosing the embryo. It does not fit the embryo snugly. The space between the amnion and the embryo, or amniotic cavity, is filled with a watery fluid, the amniotic fluid. The fluid is a cushion that protects the embryo from mechanical impacts and, at the same time, allows it freedom of movement. The chemical composition of the amniotic fluid resembles the chemical makeup of blood. Thus, during its development, the embryo is bathed by a fluid that is compatible in its chemical nature with the embryo's blood. An interesting fact is that for the embryos of birds and reptiles, as well as for the human fetus, the amniotic fluid resembles seawater in chemical composition, in terms of elements (ions), such as sodium, potassium, calcium, and magnesium. After approximately 400 million years of evolution on land, vertebrates still carry essentially the chemical composition of seawater in their fluids—amniotic fluid, blood, and other body fluids. This has been interpreted as indicating that life originated in the sea, and that the balance of salts in various body fluids did not change very

Figure 12.4 Seymouria, an advanced amphibian with reptilian body proportions (short trunk region and well-developed limbs). *Seymouria* is known only from Permian rocks (275 million years ago), long after the reptiles had appeared. It may represent a relict of an earlier amphibian group that was ancestral to the reptiles. (From a painting by Francis Lee Jaques; American Museum of Natural History.)

much in a subsequent evolution. The amnion has been picturesquely characterized as a sort of private aquarium in which the embryos of landliving vertebrates recapitulate the water-living mode of existence of their remote ancestors.

As figure 12.5 shows, the enormous mass of yolk has been enclosed by an internal circular membrane, the yolk sac. This membrane, which has formed by growing over the yolk, is attached to the embryonic body by a narrow stalk. The yolk sac is highly vascular, its many blood vessels communicating with the blood channels of the embryo proper. The blood circulating through the vessels of the yolk sac carries dissolved yolk materials to all parts of the embryo, thus making the yolk available for chemical activities and growth in all regions.

Figure 12.5   The amniotic egg. The developing embryo is enclosed by the membranous *amnion* and cushioned by amniotic fluid. The large reserve supply of food (yolk) is contained within the *yolk sac*. The sacklike *allantois* serves as a receptacle for the embryo's waste products. The outermost, vascularized enveloping membrane is the *chorion*.

From the hind region of the embryonic body, a sac bulges out on the underside and pushes its way between the yolk sac and chorion. This sac is the allantois (fig. 12.5). An appreciable part of the blood vessels lies close to the inner surface of the shell. The porous shell permits the ready exchange of respiratory gases between the external air and the internal blood. The allantoic sac serves also as a receptacle for urinary wastes. Waste fluids excreted by the embryonic kidneys pass into the cavity of the allantois. The allantois thus has both respiratory and excretory functions. Finally, there is an outermost investing membrane, the chorion, which is abundantly supplied with blood vessels. The embryo depends, in part, on this vascularized enveloping membrane for carrying on gaseous exchange with the outer air through the porous shell.

The amnion, chorion, allantois, and yolk sac in the human embryo are essentially similar to these extraembryonic membranes in the reptiles (and birds). Curiously, the yolk sac does develop in the human embryo in spite of the absence of any appreciable amount of yolk. The human yolk sac, however, remains small and functionless. There is also no elaborate development of the allantois in the human embryo; the allantoic sac never becomes more than a rudimentary tube of minute size. Nevertheless, the very appearance of the yolk sac and allantois in the human embryo is one of the strongest pieces of evidence documenting the evolutionary relationships among the widely different kinds of vertebrates. To the student of evolution this means that the mammals, including man, are descended from animals that reproduced by means of externally laid eggs rich in yolk.

Adaptive Radiation of Reptiles

With the perfection of the amniotic egg, the reptiles exploited the wide expanses of land areas. The ancestral reptilian stock initiated one of the most spectacular adaptive radiations in life's history. The reptiles endured as dominant land animals of the earth for well over 100 million years. The Mesozoic era, during which the reptiles thrived, is often referred to as the "Age of Reptiles."

Figure 12.6 reveals the variety of reptiles that blossomed from the basal stock, the cotylosaurs. The dinosaurs were by far the most awe-inspiring and famous. They reigned over the lands until the close of the Mesozoic era before suffering extinction. The dinosaurs were remarkably diverse; they varied in size, bodily form, and habits. Some of the dinosaurs were carnivorous, such as the huge *Tyrannosaurus,* whereas others were vegetarians, such as the feeble-toothed but ponderous *Brontosaurus.* The prodigious body of *Brontosaurus* weighed 30 tons and measured nearly 70 feet in length. Not all dinosaurs were immense; some were no bigger than chickens. Some dinosaurs strode on two feet; others had reverted to four. The exceedingly long necks of certain dinosaurs were adaptations for feeding on the foliage of tall coniferous trees.

The dinosaurs were descended from the thecodonts—slender, fast-running lizardlike creatures. Some authors contend that the agile thecodonts were already warm-blooded (endothermic) rather than, as often assumed, cold-blooded (ectothermic). Under this view, the dinosaurs were also active and endothermic. An endothermic animal is not dependent on external sources of heat, but has the capacity to produce and control its own internal heat. The long reign of the dinosaurs could reflect, in part, the capacity for endothermy.

The thecodonts gave rise also to bizarre reptiles that took to the air, the pterosaurs. These "dragons of the air" possessed highly expansive wings and disproportionately short bodies. The winged pterosaurs succumbed before the end of the Mesozoic era. Another independent branch of the thecodonts led to eminently more successful flyers, the birds. The origin of birds from reptiles is revealed by the celebrated *Archaeopteryx,* a Jurassic form (fig. 12.6) that was essentially an air-borne lizard. This feathered creature possessed a slender, lizardlike tail and a scaly head equipped with reptilian teeth.

Certain reptiles returned to water. The streamlined, dolphinlike ichthyosaurs and the long-necked, short-bodied plesiosaurs were marine, fish-eating reptiles. The ichthyosaurs were proficient swimmers; their limbs were finlike and their tails were forked. The plesiosaurs were efficient predators, capable of swinging their heads 40 feet from side to side and seizing fish in their long, sharp teeth. These aquatic reptiles breathed by means of lungs; they did not redevelop the gills of their very distant fish ancestors. Indeed, it

Figure 12.6 Adaptive radiation of reptiles. A vast horde of reptiles came into existence from the basal stock, the cotylosaurs, at the beginning of the Mesozoic era, roughly 225 million years ago. This matchless assemblage of reptiles was triumphant for a duration well over 100 million years. Then, before the close of the Mesozoic era, the great majority of reptiles passed into oblivion. (Adapted from Edwin H. Colbert, *The Dinosaur Book* [New York: McGraw-Hill, 1951], p. 52.)

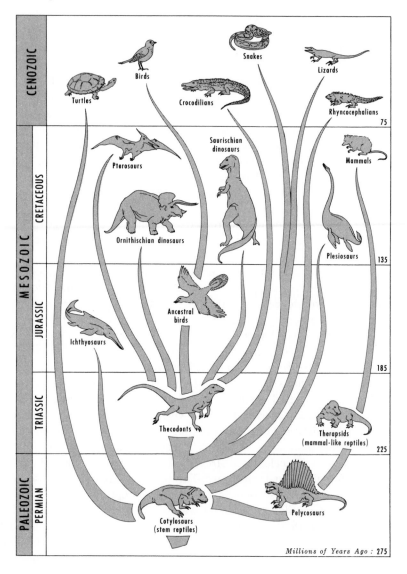

is axiomatic that a structure once lost in the long course of evolution cannot be regained. This is the doctrine of irreversibility of evolution, or *Dollo's Law,* after Louis Dollo, the eminent Belgian paleontologist to whom the principle is ascribed. The reversal of the long evolutionary path from lungs to gills would demand the impossibly precise retrieval of an untold number of steps.

Among the early reptiles present before Mesozoic days were the pelycosaurs, notable for their peculiar sail-like extensions of the back (fig. 12.7). The function of the gaudy sail is unknown, but it should not be thought that this structural feature was merely ornamental or useless. As we have emphasized, traits of organisms were selected for their adaptive utility. It may be that the pelycosauran sail was a functional device to achieve some degree of heat regulation. Be that as it may, the pelycosaurs gave rise to an important group of reptiles, the therapsids. These mammal-like forms bridged the structural gap between the reptiles and the mammals.

Figure 12.7 Dimetrodon, one of the pelycosaurs that flourished during Permian times. The gaudy "sail" may have served as a heat-regulating device. (From a restoration by Charles R. Knight; American Museum of Natural History.)

Extinction and Replacement

The history of the reptiles attests to the ultimate fate of many groups of organisms—*extinction*. The reptilian dynasty collapsed before the close of the Mesozoic era. Of the vast host of Mesozoic reptiles, relatively few have survived to modern times; the ones that have include the lizards, snakes, crocodiles, and turtles. The famed land dinosaurs, the great marine plesiosaurs and ichthyosaurs, and the flying pterosaurs all became extinct. The cause of the decline and death of the tremendous array of reptiles is obscure. The demise of the giant reptiles is generally attributed to their inability to adapt to some radical change in environmental conditions. In general, most species of organisms do become highly specialized. Perhaps many species pass unavoidably into oblivion when they cannot genetically adapt to radical environmental changes.

Whatever may be the cause of mass extinctions, the fact remains that as one group of organisms recedes and dies out completely, another group spreads and evolves. The decline of the reptiles provided evolutionary opportunities for the birds and the mammals. The vacancies in the habitats could then be occupied by these warm-blooded vertebrates. Small and inconspicuous during the Mesozoic era, the mammals arose to unquestionable dominance during the Cenozoic era, which began approximately 65 million years ago. The extent to which the mammals contributed to the decline of the dinosaur assemblage is unanswerable. Our knowledge of evolutionary history suggests that the ascendancy of the mammals was the result rather than the cause of the fall of the great reptiles.

In mammals, major evolutionary advance was the development of the placenta, an intimate apposition of maternal and fetal tissues. The placenta provides optimum conditions for the growth of the fetus—protection, a continuous supply of oxygen and food, an efficient system for removing waste products, and the transmission of antibodies. Of paramount significance, mammals have added a higher order of parental care of the young. The newborn is relatively helpless and obtains nourishment in the form of milk through special mammary glands. The most primitive mammals are the monotremes (duckbill platypus and spiny anteater), which lay reptilelike yolky eggs. Yet, the monotremes nourish their young with milk secreted by modified sweat glands.

The mammals diversified into marine forms (for example, the whale, dolphin, seal, and walrus), fossorial forms living underground (for example, the mole), flying and gliding animals (for example, the bat and flying squirrel), and cursorial types well-adapted for running (for example, the horse). The mammalian appendages are highly differentiated for the different modes of life. An important lesson may be drawn from the variety of specialized appendages. Superficially there is scant resemblance between the human arm, the flipper of a whale, and the wing of a bat. And yet, a close comparison of the skeletal elements (fig. 12.8) shows that the structural design, bone for bone, is basically the same. The differences are mainly in the relative lengths

Figure 12.8   Varied forelimbs of vertebrates, all of which are built on the same structural plan. The best explanation for the fundamentally similar framework of bones is that humans and all other vertebrates share a common ancestry. Homologous bones are indicated as follows: humerus (upper arm)—crosshatching; radius (forearm)—light stippling; ulna (forearm)—black; carpals (wrist)—white; metacarpals (palm) and phalanges (digits)—heavy stippling. The number of phalanges in each digit is indicated by a numeral, beginning with the first digit (thumb).

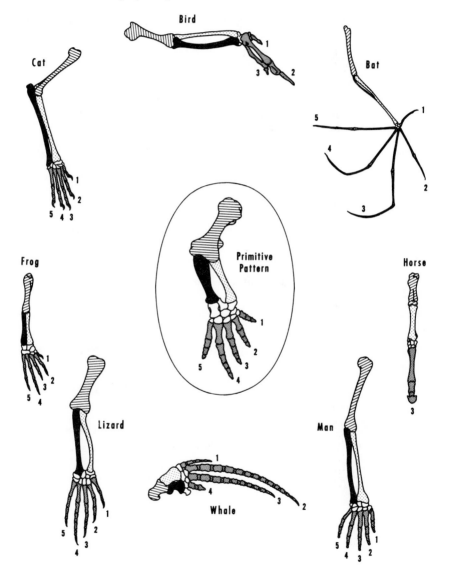

of the component bones. In the forelimb of the bat, for instance, the meta-carpals and phalanges (except those of the thumb) are greatly elongated. Al-though highly modified, the bones of the bat's wing are not fundamentally different from those of other mammals. The conclusion is inescapable that the limb bones of the human, the bat, and the whale are modifications of a common ancestral pattern. The facts admit no other logical interpretation. Indeed, as seen in figure 12.8, the forelimbs of all tetrapod vertebrates exhibit a unity of anatomical pattern intelligible only on the basis of common inheritance. The corresponding limb bones of tetrapod vertebrates are said to be *homologous,* since they are structurally identical with those in the common ancestor. In contrast, the wing of a bird and the wing of a butterfly are *analogous;* both are used for flight, but they are built on an entirely different structural plan.

Selected Readings

Colbert, E. H. 1955. *Evolution of the vertebrates.* New York: John C. Wiley & Sons.

————. 1961. *Dinosaurs, their discovery and their world.* New York: E. P. Dutton & Co.

Dobzhansky, T. 1955. *Evolution, genetics, and man.* New York: John C. Wiley & Sons.

Hotton, N. 1968. *The evidence of evolution.* New York: American Heritage Publishing Co.

Johansen, K. 1968. Air-breathing fishes. *Scientific American,* October, pp. 102–11.

Kurtén, B. 1969. Continental drift and evolution. *Scientific American,* March, pp. 54–64.

Newell, N. D. 1963. Crises in the history of life. *Scientific American,* February, pp. 76–92.

Romer, A. S. 1959. *The vertebrate story.* Chicago: University of Chicago Press.

Simpson, G. G. 1951. *Horses.* New York: Oxford University Press.

————. 1953. *Life of the past.* New Haven, Conn.: Yale University Press.

Smith, H. W. 1961. *From fish to philosopher.* Garden City, N.Y.: Doubleday & Co.

Stirton, R. A. 1959. *Time, life, and man.* New York: John C. Wiley & Sons.

Young, J. Z. 1950. *The life of vertebrates.* New York: Oxford University Press.

13
Origin of Life

The polemics surrounding Darwin's *Origin of the Species* was matched by another monumental controversy of that day centering on the unbelieving proposition that germs (bacteria) cause disease. The father of the germ theory was Louis Pasteur, the chemist who was led into the study of germs through problems associated with the fermentation of wine. In the early 19th century, the notion of spontaneous generation was still flourishing—most people believed that living things can originate from lifeless matter. In 1862, Pasteur provided proof that new life can only come from preexisting life. Following years of bitter wrangling with skeptics, Pasteur finally laid to rest the age-old belief that life could appear out of nowhere.

Since Pasteur's time, countless generations of students have been taught not to believe in spontaneous generation. However, Pasteur's experiments revealed only that life cannot arise spontaneously under conditions that exist on earth today. Conditions on the primeval earth billions of years ago were assuredly different from present conditions, and the first form of life, or self-duplicating particle, did arise spontaneously from chemical inanimate substances. It should be clear that the conditions under which life originated were unique; the origin again of life on present-day earth is scarcely possible.

Primitive Earth

The view that life emerged by a long and gradual process of chemical evolution was first convincingly set forth by the Russian biochemist Alexander I. Oparin, in 1924, in an enthralling booklet entitled *The Origin of Life*. The transformation of lifeless chemicals into living matter extended over some one billion years. Such a transformation, as Oparin points out, is no longer possible today. If by pure chance a living particle approaching that of the first form of life should now appear, it would be rapidly decomposed by the oxygen of the air or quickly destroyed by the countless microorganisms presently populating the earth.

Our earth is reliably estimated to be 4.6 billion years old. It was formerly thought that the earth originated as a fiery mass that was torn away from the sun. Astronomers now generally acknowledge that the earth (like other planets in the solar system) condensed out of a swirling cloud of gas surrounding the primitive sun. The atmosphere of the pristine earth was quite unlike our present atmosphere. Oxygen in the free gaseous state was virtually absent; it was bound in water and in metallic oxides on surface rocks and particles. Accordingly, any complex organic compound that arose during this early time would not be subject to degradation by free oxygen. Moreover, there was no layer of ozone to absorb the stark ultraviolet rays from the sun that would be lethal to modern animal life.

The early gas cloud was especially rich in hydrogen. The hydrogen (H_2) of the primordial earth chemically united with carbon to form methane (CH_4), with nitrogen to form ammonia (NH_3), and with oxygen to form water vapor (H_2O). Thus, the early atmosphere had a strongly reducing (nonoxygenic) character, containing primarily hydrogen, methane, ammonia, and water. The atmospheric water vapor condensed into drops and fell as rain; the rains eroded the rocks and washed minerals (such as chlorides and phosphates) into the seas. The stage was set for the combination of the varied chemical elements. Chemicals from the atmosphere mixed and reacted with those in the waters to form a wealth of hydrocarbons (that is, compounds of hydrogen and carbon). Water, hydrocarbons, and ammonia are the raw materials of amino acids, which, in turn, are the building blocks for the larger protein molecules. Thus, in the primitive seas, amino acids accumulated in considerable quantities and became linked together to form proteins.

Complex carbon compounds, such as amino acids and proteins, are termed *organic* because they are made by living organisms. Our present-day green plants use the energy of sunlight to synthesize organic compounds from simple molecules. What, then, was the energy source in the primitive earth, and how was synthesis of organic compounds effected in the absence of living things? It is generally held that ultraviolet rays from the sun, electrical discharges such as lightning, and intense dry heat from volcanic activity furnished the energy to join the simple carbon compounds and nitrogenous substances into amino acids. Is there a valid basis for such a widely accepted view?

Experimental Synthesis of Organic Compounds

In the early days of chemistry, it was believed that organic compounds could be produced only by living organisms. But in 1828, Friedrich Wöhler succeeded in manufacturing the organic compound *urea* under artificial conditions in the laboratory. Since Wöhler's discovery, a large variety of organic chemicals (amino acids, monosaccharides, purines, and vitamins) formerly produced only in organisms have been artificially synthesized.

In 1953, Stanley Miller, then at the University of Chicago and a student of Nobel laureate Harold Urey, synthesized organic compounds under conditions resembling the primitive atmosphere of the earth. He passed electrical sparks through a mixture of hydrogen, water, ammonia, and methane. The electrical discharges duplicated the effects of violent electrical storms in the primitive universe. In the laboratory, the four simple inorganic molecules interacted, after a mere week, to form several kinds of amino acids, among them alanine, glycine, aspartic acid, and glutamic acid. Miller's instructive experiment has been successfully repeated by a number of investigators; amino acids can also be generated by irradiating a similar mixture of gases with ultraviolet light.

The synthesis of amino acids is only a small step toward the synthesis of a living cell. In 1964, Sidney W. Fox, then at Florida State University and later at the University of Miami, reasoned that proteins were synthesized from amino acids in the primitive earth by thermal energy, or heat. He accordingly heated a mixture of 18 amino acids to temperatures of 160–200°C for varying periods of time. He obtained stable, proteinlike macromolecules, which he termed *proteinoids*. These thermally produced proteinoids are similar to natural proteins in many respects. As a striking instance, bacteria can actually utilize the proteinoids in a culture medium, degrading them enzymatically into individual amino acids. Equally important, when the proteinoid material was cooled and examined under a microscope, Fox observed small, spherical, cell-like units that had arisen from aggregations of the proteinoids. These molecular aggregates, called *microspheres,* exhibited a general resemblance to spherical bacteria. Such microspherelike aggregates could have been the forerunners of the first living organism. However, there remains a large hiatus between the formation of organic microspheres and the appearance of the first living cell, capable of specific catalytic functions, reproduction, and mutation.

Life's Beginnings

It is evident that organic compounds can be formed without the intervention of living organisms. Thus, it appears likely that the sea of the primitive earth spontaneously accumulated a rich mixture of organic molecules. In the absence of living organisms, the organic compounds would have been stable and would persist for countless years. The sea became a sort of dilute organic soup (an aquatic Garden of Eden), in which the molecules collided and reacted to form new molecules of increasing levels of complexity. Proteins capable of catalysis, or enzymatic activity, had to evolve, and nucleic acid molecules capable of self-replication must also have developed.

The living cell is an orderly system of chemical reactions (directed by enzymes) that has the ability to reproduce. It seems reasonable that the machinery for self-replication (that is, self-duplicating nucleic acids) evolved before the development of the metabolic (enzymatic) machinery of the cell. Whereas amino acids are the building blocks of proteins, nucleotides are the basic units of nucleic acids. The linear polynucleotides occur in nature in the form of deoxyribonucleic acid (DNA) or ribonucleic acid (RNA). Polynucleotides have the unique property of specifying the sequence of nucleotides in a new molecule by acting as templates for the assembly of the nucleotides. Since preferential binding occurs between pairs of nucleotides (cytosine coupled with guanine and adenine coupled with uracil), new daughter polynucleotide molecules can arise in which the nucleotide sequences are complementary to the parent molecules.

It is likely that self-replicating polynucleotides (nucleic acids) slowly became established in the primordial earth some 3.5 billion years ago. In the absence of proteinaceous enzymes in the prebiotic environment, less efficient catalysts in the form of minerals or metal ions would be sufficient to promote the reactivity of the nucleotide precursors. We may presume that errors continually occurred in the copying process, and that the sequence of nucleotides in the original polynucleotide molecule became altered on numerous occasions. Large numbers of polynucleotide variations were undoubtedly maladapted. Just as organisms today compete for available resources, molecules with different nucleotide sequences competed for the available nucleotide precursors in promoting copies of themselves. Any new mutant sequence with a replication rate higher than the antecedent sequence would be the "fittest" and prevail. Natural selection (differential reproduction) thus operates on populations of molecules as it does on populations of organisms.

Certain polynucleotides not only specified their own sequences, but apparently directed the synthesis of macromolecules of an entirely different type—namely, polypeptides (proteins). In present-day organisms, RNA guides protein synthesis, which suggests that the nucleotides of RNA (not DNA) were the first carriers of genetic information. Thus, nucleic acids established a primitive genetic code for ordering amino acids into proteins. It warrants emphasis that the genetic code—the translation of nucleotide sequences into amino acid sequences—became established at a very early stage of organic evolution. The universality of the genetic code in present-day organisms attests to the early origin of the code.

A selectively advantageous RNA molecule would be one that directs the synthesis of a protein that accelerates the replication of that particular RNA. However, a protein specified by a particular variant of RNA could not foster the reproduction of that kind of RNA unless it were restrained in the immediate vicinity of the RNA. The free diffusion of proteins could be forestalled if some form of a compartment evolved to enclose or circumscribe the specific protein made by a particular RNA. All present-day cells have a limiting (plasma) membrane composed principally of phospholipids. It is not implausible that the first cell was formed when polarized films of phospholipid formed soap-like bubbles enclosing aggregations of other complex macromolecules. Specifically, a membranous structure (plasma membrane) arose that encircled a self-replicating aggregation of RNA and protein molecules. Once bounded by a limiting membrane, a given RNA molecule could be assured of propagating its own special protein molecule.

At some later stage in the evolutionary process, DNA took the place of RNA as the repository of genetic information. In other words, the genetic code was stored in DNA rather than RNA. Today, RNA molecules serve essentially their primal function: directing protein synthesis. Thus, in modern cells, genetic information is stored in DNA, transcribed into RNA, and translated into protein. An endless permutation of nucleic acids and proteins has fostered an enormous richness of life.

We may assume that the first living systems drew upon the wealth of organic materials in the sea broth. Organisms that are nutritionally dependent on their environment for ready-made organic substances are called *heterotrophs* (Greek, *hetero,* "other" and *trophos,* "one that feeds"). The primitive one-celled heterotroph probably had little more than a few genes, a few proteins, and a limiting cell membrane. The heterotrophs multiplied rapidly in an environment with a copious supply of dissolved organic substances. However, the ancient heterotrophs could survive only as long as the existing store of organic molecules lasted. Eventually, living systems evolved the ability to synthesize their own organic requirements from simple inorganic substances. In the course of time, *autotrophs* (*auto,* meaning "self") arose, which were able to manufacture organic nutrients from simpler molecules.

The primitive autotroph would require a whole array of enzymes to direct a multistep chain of reactions involved in the synthesis of a protein. However, it would be too much to expect that all the necessary enzymes evolved at the same time. Norman H. Horowitz of the California Institute of Technology has ingeniously proposed that the chain of steps in a complicated chemical pathway evolved *backward.*

Retrograde Evolution

Horowitz proposed that an organism might acquire, by successive mutations, the enzymes required in a biosynthetic pathway in the reverse order from that in which they normally appear in a sequence. In other words, evolution began with the end product of the pathway and worked backward, one step at a time, toward the beginning of the reaction chain. The pathway became constructed in its entirety when the organism had tapped from its surroundings all of the intermediate precursors of the reaction.

Let me assume that an organic compound, *O,* is synthesized through the following steps, where *A* through *D* represent precursors:

$$A \longrightarrow B \longrightarrow C \longrightarrow D \longrightarrow O$$

The first primitive heterotroph, lacking synthetic ability, would require the presence of *O* in its environment. This essential organic chemical, continually being used by the heterotroph, would eventually become rare, if not exhausted. If a mutation occurred that endowed the heterotroph with the ability to catalyze the reaction from *D* to *O,* then the organism would no longer be dependent on the availability of O in the environment. Indeed, in an environment where O had become depleted, the new mutant would have a survival (selective) advantage over the ancestral type and replace it. As the new mutant reproduced, D in turn would become scarce. Another mutation might occur in the mutant, converting it to a still newer form capable of catalyzing the reaction C to D. At this point, both the first mutant and the newly arisen mutant could live together in a close mutual (symbiotic) relationship, establishing the first two-gene system.

The intimate two-gene combination would be able to survive and reproduce in the presence of C and D. As other compounds in synthetic pathway become progressively rarer, additional mutations for their synthesis would be favorably selected. Ultimately, a multigenic system would evolve capable of directing the synthesis of O from inorganic substance A by way of intermediate products B, C, D. At first glance, it might appear that an unreasonable number of mutational events has been invoked. But it should be recalled that mutations continually occur in living organisms. In fact, the capacity for mutation might be regarded as a cardinal property of life.

Autotrophic Existence

The first simple autotroph arose in an anaerobic world, one in which little, if any, free oxygen was available. The primitive autotrophs obtained their energy from the relatively inefficient process of fermentation (the breakdown of organic compounds in the absence of oxygen). Thus, the early fermentative autotrophs were much like our present-day anaerobic bacteria and yeast. The metabolic processes of the anaerobic autotrophs resulted in the liberation of large amounts of carbon dioxide into the atmosphere. Once this occurred, the way was paved for the evolution of organisms that could use carbon dioxide as the sole source of carbon in synthesizing organic compounds and could use sunlight as the sole source of energy. Such organisms would be the photosynthetic cells.

Early photosynthetic cells probably split hydrogen-containing compounds, such as hydrogen sulfide. In other words, as is still performed today by sulfur bacteria, hydrogen sulfide is cleaved into hydrogen and sulfur. The hydrogen is used by the cell to synthesize organic compounds, and the sulfur is released as a waste product (as evidenced by the earth's great sulfur deposits). In time, the process of photosynthesis was refined so that water served as the source of hydrogen. The result was the release of oxygen as a waste product. At this stage, free oxygen became established for the first time in the atmosphere.

The first organisms to use water as the hydrogen source in photosynthesis were the blue-green algae. Since blue-green algae were active photosynthesizers, atmospheric oxygen accumulated in increasing amounts. Oxygen would be toxic to anaerobic bacteria. Many primitive anaerobic bacteria, incapable of adapting to free oxygen, remained in portions of the environment that were anaerobic, such as sulfur springs and oxygen-free muds. New kinds of bacteria, however, arose that were capable of utilizing the free oxygen. Today, there are bacterial types that are anaerobic as well as aerobic.

The earth's atmosphere gradually changed from a reducing, or hydrogen-rich, atmosphere to an oxidizing, or oxygen-rich, atmosphere. Geologic evidence indicates that free oxygen began to accumulate in the atmosphere about two billion years ago. The rising levels of atmospheric oxygen set the stage for the appearance of one-celled eucaryotic organisms, which arose at least one billion years ago (fig. 13.1). Then, within the comparatively short span of the last 600 million years, the one-celled eucaryotes evolved in various directions to give rise to a wealth of multicellular life forms inhabiting the earth.

Evolution of Eucaryotes

The simplest cells in nature—the bacteria and blue-green algae—have been classified as *procaryotic* cells. These cells have no nuclear membrane by which the hereditary materials (DNA) are set apart from the cytoplasm, and lack specialized cytoplasmic bodies (organelles), such as mitochondria and chloroplasts. In a bacterial cell, the DNA forms a simple closed loop that is attached to the inside of the cell's membrane.

The replication of a bacterial cell occurs rapidly by the simple division of the cell into two. One bacterial cell can divide every 20 minutes, and can leave 4 billion descendants (the size of the present human population) within 11 hours. The phylogenetic relationships in the bacterial world are complex. A peculiar group of bacteria—called the archaebacteria—appears to have diverged from the true bacteria (or eubacteria) very early in evolution. The strange class of archaebacteria consists of bacteria that actually generate methane (methanogens), bacteria that thrive in salt (halophiles), and bacteria that are tolerant of acid (thermophils). Recent developments in molecular biology reveal that these bacteria share some striking biochemical features (such as a 16S ribosomal RNA sequence) that are absent from other bacteria.

In contrast to the procaryotes, the cells of the morphologically more complex plants and animals, or *eucaryotic* cells, have a distinct nuclear membrane that encloses strands of DNA, or discrete chromosomes. Eucaryotic cells also have an elaborate system of membrane-bound cytoplasmic organelles. The enzymes that are responsible for releasing energy from complex molecules are packed inside sausage-shaped organelles called mitochondria. In the eucaryotic plant cell, a prominent organelle is the chloroplast, the photosynthesizing particle that converts light energy into chemical energy. One of the most intriguing speculations is that mitochondria and chloroplasts arose not by a gradual evolutionary process but abruptly in an unusually striking manner.

Mitochondria have a number of interesting properties that suggest that they were once free-living, or independent, bacterialike organisms. Mitochondria have small amounts of their own DNA; this DNA exists as a loop-shaped

Figure 13.1   Major events in the history of life. Chemical evolution preceded biological evolution. The first self-duplicating life forms were heterotrophs. Oxygen began to accumulate in the atmosphere with the appearance of photosynthetic autotrophs. Rising levels of atmospheric oxygen were associated with the emergence of eucaryotes. The last 600 million years has witnessed a great diversity of life in an oxygen-rich atmosphere.

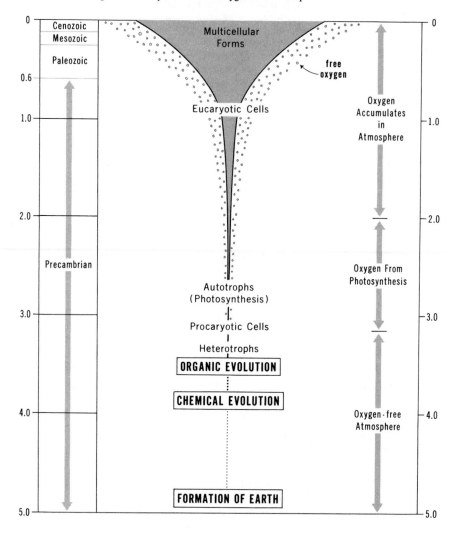

molecule like the DNA of bacteria. Mitochondria also possesses the genetic capacity to incorporate amino acids into proteins. These properties have led Lynn Margulis of Boston University to hypothesize that mitochondria may have been derived from primitive aerobic bacteria that were engulfed by predatory organisms, probably fermentative bacteria, destined to become eucaryotic (fig. 13.2). The predatory hosts became dependent on their enslaved mitochondria, and the latter, in turn, became dependent on their hosts. Thus, the association was of mutual advantage to the predatory cell and the engulfed prey. Such a close association, or partnership, is called *symbiosis*. In essence, the mitochondria (formerly oxygen-respiring bacteria) established permanent residence within the hosts. These predatory hosts became the first eucaryotic cells.

Eucaryotic plant cells contain both mitochondria and chloroplasts. Just as aerobic bacterial cells can be equated with mitochondria, blue-green algae cells are basically equivalent to the chlorophyll-containing chloroplasts. Chloroplasts, like mitochondria, have their own unique DNA and the associated protein-synthesizing machinery. We can now speculate that those predatory cells that engulfed but did not digest the aerobic bacterial cells became the eucaryotic animal cells. Those predators that captured both he photosynthetic blue-green algal cells and aerobic bacteria evolved into eucaryotic plant cells (fig. 13.2). Accordingly, the engulfed prey became permanent symbiotic residents—either as mitochondria or chloroplasts—within the predatory cell. The idea that present-day mitochondria and chloroplasts may be descendants of ancestral aerobic bacteria and blue-green algae, respectively, is highly speculative but provocative.

Some authors regard the relatively abrupt origin of the organelles of eucaryotes by a process of symbiosis as improbable. The evolution of living organisms normally proceeds through a series of gradual changes. Continual modification of the surface membrane may have been the basic evolutionary mechanism in the differentiation of eucaryotic cell from procaryotic cell. The organelles of eucaryotic cells may have evolved by the invagination, or drawing inward, of the surface membrane of a primitive cell. Figure 13.3 shows the postulated mechanism for the origin of a mitochondrion from the cell surface membrane. In support of this scheme, it is interesting that the enzymes for the breakdown of carbohydrates for energy production in bacteria are incorporated in the structure of the cell surface membrane. In like manner, other specialized internal structures of eucaryotes, including the membrane-enclosed nucleus, may have differentiated from invaginated cell surfaces (fig. 13.3). This hypothesis is simpler than the origin of eucaryotic organelles by symbiosis, and is considered more plausible by several investigators.

Figure 13.2   **Evolution of eucaryotes** through symbiosis. The eucaryotic animal cell may have arisen through a symbiotic relationship between a procaryotic cell and an aerobic bacterium. The eucaryotic plant cell may have originated by a comparable symbiotic relationship between a blue-green alga and a eucaryotic animal cell.

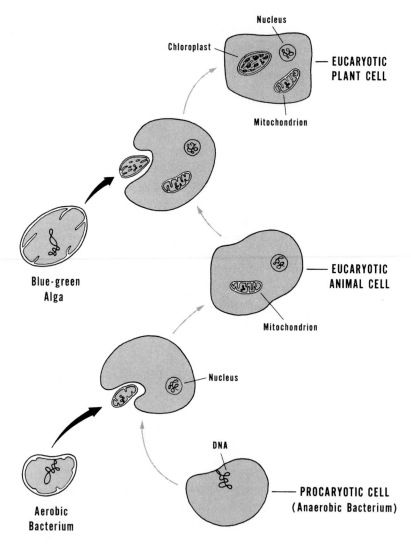

Figure 13.3 **Hypothetical scheme** of the development of the mitochondrion from the surface membrane (*A*), and of the origin of a eucaryotic cell by the invagination, or drawing inward, of the surface membrane in several places (*B*).

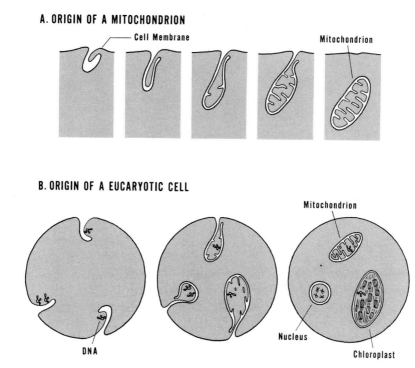

Selected Readings

Adler, I. 1957. *How life began.* New York: The New American Library.

Ayala, F. J. 1976. *Molecular evolution.* Sunderland, Mass.: Sinauer Associates, Inc.

Calvin, M. 1969. *Chemical evolution.* New York: Oxford University Press.

Carlile, M. 1982. Prokaryotes and eukaryotes: strategies and successes. *Trends in Biochemical Sciences* 7:128–30.

Cavalier-Smith, T. 1975. The origin of nuclei and of eukaryotic cells. *Nature* 256:463–68.

Cohen, S. S. 1970. Are/were mitochondria and chloroplasts microorganisms? *American Scientist* 58:281–89.

————. 1973. Mitochondria and chloroplasts revisited. *American Scientist* 61:437–46.

Dickerson, R. E. 1978. Chemical evolution and the origin of life. *Scientific American,* September, pp. 70–86.

Dickerson, R. E., and Geis, I. 1976. *Chemistry, matter, and the universe.* New York: W. A. Benjamin.

Doolittle, W. F. 1980. Revolutionary concepts in evolutionary cell biology. *Trends in Biochemical Sciences* 5:146–49.

Eigen, M.; Gardiner, W.; Schuster, P.; and Winkler-Oswatitsh, R. 1981. The origin of genetic information. *Scientific American,* April, pp. 88–118.

Fox, G. E.; Stackebrandt, E.; Hespell, R. B., *et al.* 1980. The phylogeny of prokaryotes. *Science* 209:457–63.

Margulis, L. 1970. *Origin of eucaryotic cells.* New Haven: Yale University Press.

Miller, S. L. 1953. A production of amino acids under possible primitive earth conditions. *Science* 117:528–29.

Oparin, A. I. 1953. *The origin of life.* New York: Dover Publications.

Raff, R. A., and Mahler, H. R. 1972. The nonsymbiotic origin of mitochondria. *Science* 177:575–82.

Schopf, J. W. 1978. The evolution of the earliest cells. *Scientific American,* September, pp. 110–38.

Uzell, T., and Spolsky, C. 1974. Mitochondria and plastids as endosymbionts: A revival of special creation? *American Scientist* 62:334–43.

Woese, C. R. 1971. Archaebacteria. *Scientific American,* June, pp. 98–122.

Woese, C. R., and Fox, G. E. 1977. Phylogenetic structure of the prokaryotic domain: The primary kingdoms. *Proceedings of the National Academy of Sciences* 74:5088–90.

14
Molecular Evolution

One of the dramatic breakthroughs in the 1950s was an understanding of the interrelations of protein structure and the genetic information carried in deoxyribonucleic acid (DNA). There emerged the fundamental tenet that the gene (a linear array of nitrogenous bases in the DNA) codes for the precise sequence of amino acids that compose a protein. There was also the realization that gene mutation often involves no more than an alteration of a single base in the DNA sequence that specifies the amino acid. Such a single base alteration can result in the substitution of one amino acid for another in the protein molecule, which interferes with the activity of the protein. This is vividly exemplified by the aberrant hemoglobin of sickle cell anemia, in which a highly localized change in one of the bases of the DNA molecule leads to the substitution of only one amino acid residue in the hemoglobin molecule.

Over evolutionary time, DNA sequences diverge and proteins change in composition. The amino acid sequences of proteins have gradually been modified with time, yielding the arrays that are found today in extant species. Comparisons of the amino acid compositions in present-day organisms enable us to infer the molecular events that occurred in the past. The more distant in the past that an ancestral stock diverged into two present-day species, the more changes will be evident in the amino acid sequences of the proteins of the two contemporary species. Viewed another way, the number of amino acid modifications in the lines of descent can be used as a measure of time since the divergence of the two species from a common ancestor. In essence, protein molecules incorporate a record of their evolutionary history that can be just as informative as the fossil record.

Amino Acid Substitutions

The simplest alteration that can occur in a protein is the replacement of one amino acid for another as a consequence of a single substitution of a nitrogenous base in the DNA molecule. Each amino acid is normally coded by a group (triplet) of three nitrogenous bases. As shown in figure 14.1, the code is read linearly, starting from one end, in consecutive blocks of three bases. Accordingly, the reading of the code could be modified if an error occurred, such as the substitution, addition, or elimination (deletion) of a nitrogenous base.

A substitution of one of the DNA bases constitutes a gene (point) mutation. Some point mutations will have a profound effect on protein function if they alter the folding conformation of the peptide chains or interfere with the active sites essential for binding substrates. The greatest disturbance of function of the hemoglobin molecule, for example, occurs when the amino acid sequence is altered in that section of the polypeptide chain immediately involved in binding oxygen. As indicated in figure 14.1, a point mutation that results in the replacement of one amino acid for another is called a *missense* mutation. If the triplet coding for the amino acid changes to a terminating codon (which signals the termination of formation of the polypeptide chain), the change is called a *nonsense* mutation.

Additions or deletions of bases in the DNA molecule typically result in an altered sequence of amino acids in the protein. For example, if the base

Figure 14.1 The molecular basis of point mutation. Each substitution, deletion, or insertion is a mutational event that alters the DNA sequence and, in turn, the polypeptide (protein) product.

sequence ATGCTTCTC is normally read [ATG][CTT][CTC], then an insertion of the base C between [ATG] and [CTT] would lead to a shift in the "reading frame," as follows: [ATG][CCT][TCT][C..]. Such modifications are called *frameshift* mutations.

Some point mutations are said to be "silent" since they do not alter the amino acid sequence. Most amino acids are specified by more than one kind of DNA triplet (see chapter 7). For example, the DNA triplet TCG codes for the amino acid serine. If base G changes to A, the triplet TCA will still code for serine. In fact, the amino acid serine can be specified by six DNA triplets (or six RNA codons). One of the surprises has been the finding that the number of silent base substitutions during the evolutionary divergence of some protein molecules has been as high as 40 percent.

Evolution of Mammalian Antidiuretic Hormones

Small peptides (octapeptides) produced by the hypothalamus of the brain provide an instructive example of the evolution of both structure and function at the molecular level. One of the notable octapeptides is the antidiuretic hormone, which promotes the reabsorption of water by the kidney tubules. There are three chemical variants of this water-conserving hormone (fig. 14.2). Arginine vasopressin is the antidiuretic hormone of most mammals, including humans. The water-conserving function is served by lysine vasopressin in the pig, and by arginine vasotocin in amphibians, birds and reptiles. Arginine vasotocin also exists in bony fishes, but an antidiuretic response is not known to occur in the kidneys of bony fishes. This antidiuretic hormone thus appeared in vertebrate phylogeny before it had acquired an antidiuretic function.

The evolution of the variants of the antidiuretic hormone clearly occurred by single base changes in the DNA molecule, yielding functional variant amino acids. The mammalian arginine vasopressin represents an amino acid substitution in the third position of the molecule where phenylalanine replaces isoleucine (fig. 14.2). In the domestic pig, lysine replaces arginine in position 8 of the molecule.

Oxytocin is an hypothalamic octapeptide that promotes the flow of milk in the mammary glands. Although vastly different functionally from vasopressin, oxytocin represents but a single amino substitution in the vasopressin molecule. Thus, by a mere change in one amino acid, a marked alteration in function has occurred. Oxytocin is an essential hormone in mammalian lactation and parturition, and is not known to play any role in the reproductive processes of nonmammalian vertebrates. Yet, oxytocin evolved in the fishes millions of years ago. Whatever the role in fishes, oxytocin serves as an example of an ancient molecule adapted to a new use.

The origin of oxytocin reveals that the evolutionary process is both conservative and opportunistic. A new product in evolution often arises not by *de nova* production but by the modification of a preexisting product.

Figure 14.2 Phylogenetic relationships of octapeptide hormones.

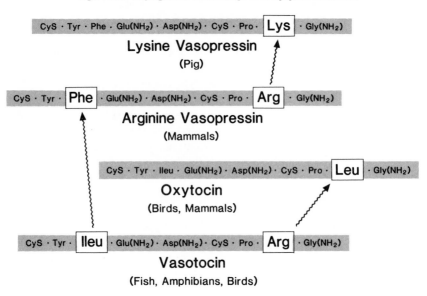

Cytochrome *c*

Cytochrome *c* is an ancient, evolutionarily conservative molecule that serves as an essential enzyme in respiration. It is a relatively small protein of a chain length slightly over 100 amino acid residues, which apparently exists in all eucaryotic organisms. This ubiquitous protein has been extracted, purified, and analyzed in more than 80 eucaryotic species. It has 104 amino acids in vertebrates and a few more in certain species lower in the phylogenetic tree. Since cytochrome *c* performs similar vital functions in all organisms, fairly stringent evolutionary constraints have been placed on the acceptance of chance alterations of the primary amino acid sequence.

Some amino acid residues of cytochrome *c* have not varied at all throughout time. In particular, the same amino acids have been found in 20 positions in all organisms tested, from molds to humans. Apparently, the 20 amino acids at specific positions are irreplaceable, and any substitutions at these sites are likely to interfere with the function of the enzyme. Indeed, there is one region of cytochrome *c* formed by a sequence of 11 amino acids (positions 70 through 80) that has remained intact in all organisms.

The number of amino acid replacements in cytochrome *c* of several species are compared in table 14.1. First, it is evident that the chromosomes of closely related vertebrates differ in only a few residues, or not at all. There is no difference at all in the composition of the amino acid residues between humans and the chimpanzee. Secondly, the greater the phylogenetic differences, the greater the likelihood that the cytochrome compositions differ. This is vividly apparent when one compares a vertebrate with an insect (tuna and moth), or an insect with yeast.

Table 14.1
Evolution of cytochrome *c*

A. Comparison between Different Organisms

Organisms	Number of Variant Amino Acid Residues
Cow and sheep	0
Cow and whale	2
Horse and cow	3
Rabbit and pig	4
Horse and rabbit	5
Whale and kangaroo	6
Rabbit and pigeon	7
Shark and tuna	19
Tuna and fruitfly	21
Tuna and moth	28
Yeast and mold	38
Wheat germ and yeast	40
Moth and yeast	44

B. Comparison with Human Cytochrome *c*

Organisms	Number of Differences of Amino Acid Residues
Chimpanzee	0
Rhesus monkey	1
Rabbit	9
Cow	10
Kangaroo	10
Duck	11
Pigeon	12
Rattlesnake	14
Bullfrog	18
Tuna	20
Fruitfly	24
Moth	26
Wheat germ	37
Mold (*Neurospora*)	40
Baker's yeast	42

*Compiled from studies by R. E. Dickerson.

How does one interpret the variations in amino acid sequences? Some of the substitutions could signify that amino acid replacements at some positions are irrelevant for the function of the protein. That is to say, amino acid substitutions at certain positions are not likely to provide either an evolutionary advantage or disadvantage, and could be preserved by chance. In essence, it may be that certain amino acid replacements are not fostered by natural selection but result from neutral mutations that are fixed by genetic drift. The notion that amino acid substitutions may be fixed by the random drift of neutral mutations remains controversial. Those who oppose the neutralist position ascribe all amino acid substitutions to natural selection. The selectionists assert that each amino acid in a protein has been selected because the given amino acid confers an advantage to the organism.

Those who adhere to the neutralist position point with satisfaction to the analysis of the fibrinopeptides, two 20-residue fragments (A and B) that are cleaved from the protein fibrinogen when it is activated to form fibrin in a blood clot. Among several mammalian species, there is an extraordinarily high level of amino acid substitutions in fibrinopeptide A. This peptide presumably functions equally well with numerous different amino acid sequences. Indeed, virtually any amino acid change appears to have been acceptable. It has been argued that the mutations responsible for the many different amino acids must be neutral. Since the fibrinopeptides have little known function except to permit the activation of fibrinogens by being discarded, they apparently can tolerate almost any amino acid change.

In other cases, however, it is difficult to deny the strong role of natural selection. Certain histones that bind to DNA in the nucleus have been highly conserved in their amino acid sequences. For example, histone IV from such divergent organisms as the pea plant and the calf differ in only 2 of 102 amino acid residues. All other amino acid substitutions apparently would disrupt the activity of the histone IV, and such substitutions that may have appeared in the evolutionary past were most likely eliminated by natural selection. Since histone IV plays a vital role in the expression of the hereditary information encoded in DNA, it is hardly surprising that this protein would be closely specified and serutinized.

Hemoglobin

Human hemoglobin has provided an ideal model for inferring the molecular events that have occurred during evolution. The oxygen-carrying molecule of the red blood cell is made up of four polypeptide chains. The hemoglobin molecule of an adult human is composed of two identical alpha chains and two identical beta chains. Each alpha chain is a linear array of 141 amino acid residues; each beta chain has 146 amino acid residues. In the amino acid sequence of the alpha chain, there is only one difference in the composition of the amino acid residues between humans and the gorilla. This contrasts sharply with differences of 19 amino acid residues in the alpha chain between humans and the pig, or differences of 26 amino acid units between humans and the rabbit. The striking molecular similarity in the hemoglobin of humans and apes holds up as well for the beta chain.

The syntheses of the alpha and beta chains of normal adult hemoglobin (designated Hb A) are specified by two pairs of genes at different loci. This state of affairs is shown in figure 14.3, where α and β denote the respective genes that control the syntheses of the polypeptide chains. The illustration also reveals that adult hemoglobin is made up of two fractions: hemoglobin A (Hb A), which is the largest component, and hemoglobin A_2 (Hb A_2), which constitutes about 2.5 percent. Hemoglobin A_2 consists of two alpha chains, which are identical to those in hemoglobin A, and two delta chains.

A different hemoglobin is present during fetal life. The alpha chain of fetal hemoglobin (Hb F) is chemically identical to the alpha chain of adult hemoglobin (Hb A) and under control by the same gene. However, in place of the beta chain, fetal hemoglobin contains a chemically different gamma chain that is governed by a separate gene (γ). Fetal hemoglobin allows increased amounts of oxygen to be transported to the tissues of the fetus. Fetal hemoglobin carries as much as 30 percent more oxygen at low tensions than does adult hemoglobin. The level of Hb F is high during the last trimester of pregnancy, but then falls off gradually until there is virtually none six months postnatally. As the synthesis of the gamma chain decreases, the synthesis of the beta chain takes over.

The locus (δ) controlling the delta chain is closely adjacent to the locus (β) governing the beta chain. This has led to the thesis that one locus arose as a copy of the other through the process of gene duplication and mutational changes. A redundant copy of a functional gene derived by gene duplication can escape scrutiny by natural selection. While sheltered from selection, the duplicated locus can accumulate mutations that were formerly rejected and emerge as a new gene with a different function than the original gene. As discussed below, duplicated genes are the raw materials of molecular evolution.

Gene Duplication

The process by which genes have evolved through duplication, followed by mutation and selection, is depicted in figure 14.4. When a particular locus duplicates itself, two genes with identical base sequences coexist in the same individual. Inasmuch as only one locus is necessary to produce the polypeptide chain, an additional locus might be regarded as superfluous. On the contrary, the duplicated locus might prove valuable in protecting the individual against the hazards of mutational changes. Ordinarily a mutation that interferes adversely with the functioning of a locus would be detrimental or lethal. However, an otherwise lethal mutation could be accepted if it occurred in the duplicated locus, since the original gene is unimpaired and would continue to synthesize the essential polypeptide. With the continual occurrence of mutations, the duplicated locus would produce a polypeptide whose sequence of amino acids is different from the original polypeptide. This new polypeptide, although still related to the original one, would be sufficiently altered to have a new and different activity and specificity. In this manner, two polypeptide chains, each with its own gene, would ultimately evolve from a single polypeptide and a single gene.

Figure 14.3 Genetic control of hemoglobin synthesis by four loci. The beta locus and the delta locus are closely linked in the same chromosome.

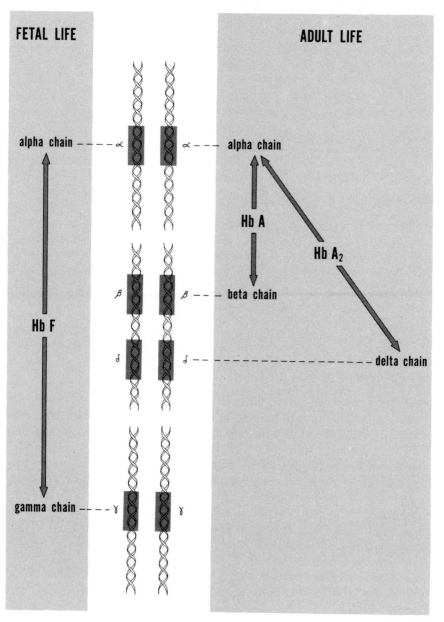

Figure 14.4 **Duplication of a gene** followed by random mutations and natural selection can lead to two genes with different functions. In the absence of gene duplication, a mutation would be lethal if the mutant gene produces an aberrant protein *(left.)* With a duplicated gene, the same mutation that would otherwise be lethal is not selectively disadvantageous, since the original gene continues to produce the normal protein *(right).* With time, several mutations can accumulate in the duplicated gene.

SELECTIVE ADVANTAGE OF GENE DUPLICATION

Normal Gene   Normal Gene

Mutation Only   Gene Duplication and Mutation

Mutation   Mutation

Abnormal Protein   Normal Protein   Abnormal Protein (Inconsequential)

Additional Mutations and Selection

Selection Against Defective Gene   Two Genes With Different Functions

A particular gene may be duplicated more than once, and large families of related proteins may eventually evolve from a common progenitor. Indeed, it is generally accepted that the varied hemoglobin genes, α, β, γ, and δ, arose by duplication from a single common ancestral gene. Each duplication was followed by the accumulation of mutations in the duplicated locus. The evolution of different molecular species of a protein thus involved the interplay of gene duplication and mutation. Duplications that originated early in the evolutionary history were frequently separated into different chromosomes by translocations, whereas recently duplicated genes remained linked on the same chromosome. The α and β genes of hemoglobin presently reside in separate chromosomes, whereas the more recently evolved δ gene has remained closely adjacent to the β gene.

Evolutionary History of Hemoglobin

We have thus far suggested that the genes involved in the syntheses of the polypeptide chains of hemoglobin were derived through duplications of a common ancestral locus, followed by divergent mutations. The original hemoglobin molecule probably resembled the chemically related myoglobin molecule. Myoglobin does not function as a respiratory protein; it is used mainly for the storage of oxygen in metabolically active tissues, especially within muscles. The storage of oxygen is very important in dividing animals, which have exceptionally high concentrations of myoglobin in their muscles. Mammalian myoglobin is a single polypeptide chain of 153 amino acid residues.

In the distant past, it is likely that myoglobin in tissue cells and hemoglobin in blood cells were identical. Myoglobin evolved into a storage protein with a high oxygen affinity, and hemoglobin became progressively adapted to fulfill its primary role of oxygen transport in different species. Based on the degree of similarity (or homology) of the amino acid residues between any two polypeptide chains, an evolutionary tree can be constructed (fig. 14.5).

Figure 14.5 Molecular evolution of hemoglobin chains. The small circles at the foci of branching indicate where the ancestral genes were duplicated to give rise to new globin chains.

ANCESTRAL CHAIN

1.0 BILLION YEARS AGO

500 MILLION

200 MILLION

40 MILLION

Myoglobin   Alpha   Gamma   Beta   Delta

Each branching point in the tree indicates where an ancestral locus was duplicated, giving rise each time to a new gene line.

It is possible to establish a time scale for the evolution of the hemoglobin chains. Nobel laureate Linus Pauling has calculated that one amino acid substitution has occurred every 7 million years. The numerous differences in the amino acid sequences between myoglobin and hemoglobin indicate that the two molecules had a common origin at least 650 million years ago, well before the first appearance of the vertebrates. A one-chain hemoglobin still persists in primitive fish, the cyclostomes. While the cyclostomes (lamprey and hagfishes) have a single hemoglobin chain, the carp (bony fish) has two chains. Accordingly, the differentiation of the alpha and beta chains of hemoglobin occurred about 500 million years ago, when the first bony fishes appeared. At that time, the alpha and beta chains became aggregated to produce the efficient four-chain hemoglobin molecule characteristic of most vertebrates.

During mammalian evolution, some 200 million years ago, the β locus duplicated once again to give rise to the γ gene that codes for fetal hemoglobin with its additional oxygen-binding capacity. The most recent duplication resulting in the δ gene occurred only 40 million years ago. The delta chain has been found only in higher primates. The divergence of the δ gene occurred prior to the separation of the phylogenetic lineage leading to the Old World monkey assemblage and the line represented by Old World monkeys, great apes, and humans. Curiously, present-day Old World monkeys do not synthesize delta chains, at least in detectable amounts. Apparently, the δ gene became silenced recently in Old World monkeys.

Pseudogenes

Genetic studies have established that the α gene and the β gene reside in different chromosomes. Indeed, a cluster of α-like genes has evolved from a single ancestral α gene by a series of duplications in one chromosome, and a comparable family of β genes has emerged in another chromosome. As shown in figure 14.6, the linked group of α-like genes on human chromosome 16 contains two active embryonic *zeta* (ζ) and two active alpha (α) genes. The β-like cluster on human chromosome 11 comprises five active genes—namely, one *epsilon* (ϵ), two *gamma* (γ), one *delta* (δ), and one *beta* (β).

Both families of hemoglobin genes contain loci called "pseudogenes" (depicted as *psi, ψ*) that cannot encode a functional polypeptide chain. Although each pseudogene shares many base sequences with its corresponding normal gene, the presence of "frameshift" mutations (see fig. 14.1) has shifted the triplets of bases so that polypeptide synthesis is prematurely halted. Pseudogenes apparently are products of gene duplication that have accumulated debilitating base changes during sequence divergence. If the $\psi\alpha 1$ gene ever becomes functional, then the alpha chain would be coded by three operational α genes.

Figure 14.6 The two families of human hemoglobin genes form clusters on different chromosomes. The *alpha* cluster of linked genes occupies a length of about 30 Kilobases (30,000 base pairs). The *beta* cluster of genes is spread over a region of about 50 Kb (50,000 base pairs.)

HUMAN HEMOGLOBIN GENES

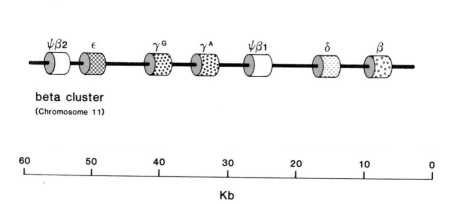

The occurrence of two immediately adjacent genes, both of which are expressed, is a common feature of gene clusters. Thus, two α genes (α1 and α2) are expressed in the production of alpha polypeptides. Such duplication of virtually identical genes has proved to be providential to some patients afflicted with the blood disorder, *thalassemia*, characterized by a reduced rate of alpha chain synthesis as a consequence of a defective α gene. Some synthesis of alpha chains becomes possible if the impairment involves only one of the several α loci.

The genes in both hemoglobin linkage groups are separated by substantial tracts of intergenic DNA. In other words, only about 10 percent of the DNA in these clusters actually code for hemoglobin polypeptides. The function of the remaining 90 percent of intergenic DNA between the genes is presently enigmatic. Some investigators have suggested that these extensive intergenic regions might have no function, and actually represent "junk" DNA. If such DNA is meaningless, then the base sequences could change rapidly during evolution without consequences. In this vein, it is of interest that the arrangements of the entire β cluster of genes are indistinguishable in the gorilla, baboon, and humans. This indicates that the arrangement of the human cluster has been fully established 20 to 40 million years ago, and has been faithfully preserved since. This conservation of the arrangement suggests that the base sequences throughout the β region have been substantially constrained by selection, at least during recent primate evolution. Such conservative evolution is incompatible with the notion that intergenic DNA is "junk."

Multiple Forms of Enzymes

The capacity of polypeptide chains to aggregate into *multimers* was an important event in protein evolution. Myoglobin, for example, exists as a monomer, but the alpha and beta chains of hemoglobin (derived from the myoglobin-like precursors) are able to form tetramers. Great functional subtlety is introduced by the formation of mixed multimers. This is well illustrated by the enzyme lactate dehydrogenase (LDH). This enzyme is indispensable to life by virtue of its metabolic role in presiding over the catalytic interconversion of pyruvate and lactate. Pyruvate is a strategically placed intermediate product in the oxidation of glucose. In the absence of oxygen, the metabolic breakdown of glucose would soon cease if it were not for the presence of lactate dehydrogenase. This enzyme makes it possible for cells to continue to function effectively during transient deficiencies in the availability of oxygen.

Since lactate dehydrogenase is so fundamental to the metabolism of the cell, one might have expected that selection would have strictly conserved the integrity of the amino acid sequences that arose initially. However, experimental analyses have shown that LDH occurs in multiple molecular forms (as isozymes) in different species and, in fact, in different tissues of the same organism. LDH is a tetramer, in which the two subunits or monomers, H and M, formed under the direction of two corresponding genes, can combine in several ways (fig. 14.7). All five possible combinations of the subunits have been found in various tissues.

The subunits in humans are designated M and H, for muscle and heart, respectively, since the lactate dehydrogenase of skeletal muscle is mainly M_4, whereas that of the heart is largely H_4. The subunit H minimizes, or at least does not promote, the reaction from pyruvate to lactate; hence, the reaction in heart muscles tends away from lactate toward pyruvate. Contrarily, the subunit M encourages the accumulation of lactate from pyruvate. This represents an exceptional instance of fine tuning by the evolutionary processes. The effects of the two subunits of LDH enable two highly specialized tissues in the body to act appropriately—heart muscle can metabolize carbohydrate efficiently only under continual aerobic conditions, and skeletal muscle can function efficiently even with the anaerobic breakdown of carbohydrate, especially under conditions of oxygen scarcity.

Figure 14.7 Formation of mixed multimers. Two polypeptide chains, if allowed to aggregate randomly, will give rise to five types of tetramers; two "pure" and three "mixed." The lactate dehydrogenase isoenzymes are formed in this manner. Isoenzymes are proteins with different structures that catalyse identical reactions (also called isozymes). Skeletal muscle largely contains the pure multimer M_4, whereas heart muscle is mainly the pure multimeter H_4.

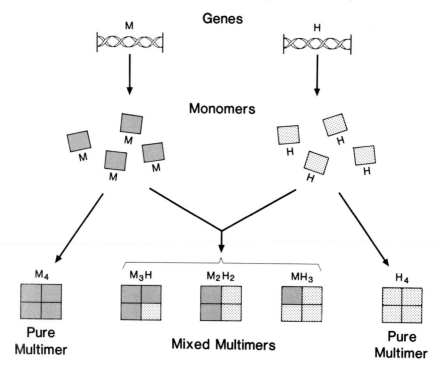

Regulating Genes

Molecular geneticists have defined two classes of genes: (1) *structural* genes, which are responsible for the actual synthesis of specific enzymatic proteins and structural cellular proteins, and (2) *regulatory* genes, which function in regulating the structural genes. Each structural gene has at least one regulatory gene and may even have several regulatory genes controlling its activity. The structural genes may be said to have a *qualitative* action, in that they transcribe information about kinds and sequences of amino acids to be incorporated in cellular proteins. On the other hand, the effect of the regulatory genes on the production of cellular proteins is *quantitative*. That is to say, the regulatory genes determine when the production of a protein will be repressed, and thereby affect the rate of synthesis of the constituent amino acids. Any change in the function of the regulatory genes would therefore result not in a change in the sequence of amino acids in a protein, but in the rate of synthesis.

When loci are duplicated, the duplicated DNA sequences need not evolve into structural genes. Indeed, during the course of evolution, several, or even many, of the duplicated loci may have become new regulatory genes. This viewpoint is plausible in light of the relatively large amounts of DNA in mammalian cells. A mouse cell, for example, may contain up to 1,000 times as much DNA as a bacterial cell, yet both types of cells contain nearly equivalent numbers of structural genes to carry out similar vital processes. In other words, the DNA of the mouse could not possibly be subdivided solely into different structural genes in single copies. In fact, in the mouse, there are many copies of several of the DNA sequences, each often several hundred nucleotides in length. The quantities of DNA involved in the repetitious sequences may constitute 30 percent of the total DNA content. It seems improbable that the repeated DNA sequences are all structural genes that code for proteins. In other words, it is highly likely that the repetitious DNA is involved in gene regulation rather than protein production. Finally, it may be that regulatory genes have been responsible for major evolutionary changes. In this respect, we shall consider the evolutionary differences between humans and chimpanzees.

Human and Chimpanzee Genes

Chimpanzees (*Pan troglodytes*) and humans (*Homo sapiens*) are two of the most thoroughly studied species, both at the organismal and molecular levels. In recent years, a variety of protein molecules of the two species have been analyzed by amino acid sequencing, immunology, and electrophoresis. All the molecular data agree in showing that chimpanzees and humans are remarkably similar. The two species have completely identical fibrinopeptide, cytochrome *c*, and alpha, beta, and gamma chains of hemoglobin. With respect to myoglobin and the delta chain of hemoglobin, only a single amino acid replacement separates the human polypeptide chain from that of the chimpanzee. In essence, the average human polypeptide is less than 1 percent different from its chimpanzee counterpart. The exceedingly meager molecular difference is much too small to account for the substantial anatomical and behavioral differences between the two species.

It has been suggested that major organismal differences, anatomically and behaviorally, are more often based on changes in regulatory systems controlling the expression of structural genes than on changes in the structural genes themselves. In particular, the complex biological differences between humans and chimpanzees stem primarily from mutational changes in regulatory genes. The mutational changes include chromosomal rearrangements (inversions and translocations) that can shift a given regulatory gene from its normal position to a new location in another chromosome. It is striking that the banding patterns of the chromosomes of humans and chimpanzees differ appreciably. This suggests considerable differences in the arrangement of genes in the chromosome complements of the two species. It may be, then, that complex anatomical and behavioral evolution is largely associated with changes in gene regulation brought about by chromosomal rearrangements. If regulatory genes are important in evolutionary changes, then *how* a protein is produced *quantitatively* is as important as *what* protein is produced *qualitatively*.

Selected Readings

Ayala, F. J. 1976. *Molecular evolution.* Sunderland, Mass.: Sinauer Associates, Inc.

Dickerson, R. E. 1972. The structure and history of ancient protein. *Scientific American,* April, pp. 58–72.

————. 1971. The structure of cytochrome *c* and the rates of molecular evolution. *Journal of Molecular Evolution* 1:26–45.

Dobzhansky, T.; Ayala, F. J.; Stebbins, G. L.; and Valentine, J. W. 1977. *Evolution.* San Francisco: W. H. Freeman and Co.

Doolittle, R. 1976. The evolution of vertebrate fibrinogen. *Federation Proceedings* 35:2145–48.

————, Wooding, G. L.; Lin, Y.; and Riley, M. 1971. Hominoid evolution as judged by fibrinopeptide structures. *Journal of Molecular Evolution* 1:74–83.

Efstratiadis, A.; Posakony, J. W.; Maniatis, T.; *et al.* 1980. The structure and evolution of the human β-globin gene family. *Cell* 21:653–68.

Fitch, W. M., and Langley, C. H. 1976. Protein evolution and the molecular clock. *Federation Proceedings* 35:2092–97.

————, and Margoliash, E. 1967. Construction of pylogenetic trees. *Science* 155:279–84.

Fox, S. W., and Dose, K. 1972. *Molecular evolution and the origin of life.* San Francisco: W. H. Freeman and Co.

Goodman, M.; Moore, G. W.; and Matsuda, G. 1975. Darwinian evolution in the genealogy of haemoglobin. *Nature* 253:603–8.

Jukes, T. H. 1968. *Molecules and evolution.* New York: Columbia University Press.

————. 1980. Silent nucleotide substitutions and the molecular evolutionary clock. *Science* 210:973–78.

Lewontin, R. C. 1974. *The genetic basis of evolutionary change.* New York: Columbia University Press.

Oparin, A. I. 1953. *The origin of life.* New York: Dover Publications.

Powell, J. R. 1975. Protein variation in natural populations of animals. *Evolutionary Biology* 8:79–120.

15
History of Life

Our interpretation of the ever-changing procession of life is based in great measure on the fossils of animals and plants entombed in the earth's rocks. An organism becomes preserved as a fossil when it is trapped in soft sediment that settles at the bottom of a lake or ocean. The deposits of mud and sand harden into the sedimentary, or stratified, rocks of the earth's crust. Fossils, then, are the remains of past organisms preserved or imprinted in sedimentary rocks. Not all fossils are mere impressions of buried parts. Exceptionally, the whole organism may be preserved, as attested by the woolly mammoths embedded in the frozen gravels of Siberia. In the case of petrified tree trunks, the wood had been infiltrated with mineral substances that crystallized and hardened. Incredibly, such mineral-impregnated wood can be sliced in very thin sections and studied microscopically for details of cellular structures.

The sample of past life is incomplete and uneven. Innumerable organisms with soft tissues (jellyfishes and flatworms, for example) are not favorable for preservation. In contrast, the woody parts of plants, the shells of mollusks, and the bones of vertebrates are relatively common as fossils. In many places of the earth, the sedimentary rocks have been subjected to such pressure and heat that their fossil holdings have been irretrievably lost. The earth's rocks have also been exposed to the relentless forces of weathering and erosion. Despite the destructive work of geologic processes, the available fossil assemblage contains an extraordinary amount of information. The older strata of rock, those deposited first, bear only relatively simple kinds of life, whereas the newer, or younger, beds contain progressively more and more complex types of life. The increasing complexity of fossils in progressively more recent rock strata is one of the more profound evidences for evolutionary changes in organisms with time. Each species of organisms now living on the earth has developed from an earlier ancestral form.

Table 15.1
Geologic Time Scale

Era	Period	Epoch
	Quaternary	Recent
		Pleistocene
Cenozoic (Age of Mammals)	Tertiary	Pliocene
		Miocene
		Oligocene
		Eocene
		Paleocene
	Cretaceous	
Mesozoic (Age of Reptiles)	Jurassic	
	Triassic	
	Permian	
	Carboniferous*	
Paleozoic	Devonian	
	Silurian	
	Ordovician	
	Cambrian	
Precambrian		

*The early Carboniferous is often referred to as the "Mississippian" and the late Carboniferous as the "Pennsylvanian."

Geologic Ages

The arrangement of geologic time is recorded in table 15.1. The main divisions, or *eras,* embrace a number of subdivisions, or *periods,* and these periods are further refined into *epochs.* Fossils appear in abundance in the rocks of the Cambrian period, which dawned 600 million years ago. The Cambrian strata bear numerous fossils of marine plants and animals; seaweeds, sponges, shellfishes, starfishes, and bizarre arthropods called trilobites.

Millions of Years Ago	Important Events
0.01	Modern humans
2	Early humans
6	Radiation of apes
23	Abundant grazing mammals
38	Angiosperms dominant
54	Mammalian radiation
65	First placental mammals
135	Climax of reptiles, first angiosperms, extinction of ammonoids
180	Reptiles dominant, first birds, first mammals
225	First dinosaurs, cycads and conifers dominant
275	Widespread extinction of marine invertebrates, expansion of primitive reptiles
345	Great swamp trees (coal forests); amphibians prominent
395	Age of fishes, first amphibians
435	First land plants; eurypterids prominent
500	Earliest known fishes
600	Abundant marine invertebrates, trilobites and brachiopods dominant, algae prominent
> 3000	Soft-bodied primitive life

Cambrian life did not come into existence abruptly. The evolutionary epic began at a far more distant time. At least 3 billion years of slow organic evolution preceded the diversity of organisms of the Cambrian period. The Precambrian seas swarmed with a great variety of microscopic forms of life. Studies in recent years have dispelled the notion that Precambrian rocks are devoid of fossils. The electron microscope has been used to examine samples of deep sedimentary rock from South Africa (the so-called Fig Tree sediments). The microscopic analysis uncovered impressions of rod-shaped bacteria that existed 3.2 billion years ago. In addition, minute traces of two chemical substances, phytane and pristane, have been found in the Fig Tree rocks. These chemicals are the relatively stable breakdown products of chlorophyll. Thus, photosynthesis by algae, in which oxygen is released, may have begun about 3 billion years ago. Other ancient rocks—such as the 2-billion-year-old Gunflint rock formation in Ontario, Canada—contain an assemblage of fossil microorganisms that appear to be blue-green algae, possibly red algae, and even some fungi. Eucaryotic organization is clearly in evidence in deposits that are approximately one billion years old. The Bitter Springs formation of Australia contains eucaryotic fossils that represent higher algae and fungi. The transition from procaryotes to eucaryotes apparently occurred during the late Precambrian era.

We may visualize the vast span of time by compressing the history of life into 60 minutes. In the 60-minute clock depicted in figure 15.1, each second represents one million years. The Precambrian era is conservatively viewed as originating at 12 o'clock, or 3.6 billion years ago. Primitive bacteria have left imprints in the sediments 3.2 billion years ago, or about 7 minutes after the start of the clock. The Precambrian era stretches over five-sixths of the earth's entire history, during which time primitive life was confined to the sea.

Toward the close of the Precambrian 700 million years ago (or 11 minutes and 40 seconds before the hour), the fossil-bearing rocks show traces of soft-bodied invertebrates of simple organization. At the onset of the Paleozoic era 600 million years ago, the fossil record improves immeasurably. The Paleozoic era begins with only 10 minutes left in the hour and occupies a span of only 6 minutes and 15 seconds (or 375 million years). Nevertheless, within this short interval, we witness the invasion of land by plants, the emergence of great swamp trees, and the dominance of fishes and amphibians. The Mesozoic era, during which the reptiles reigned, commences with less than 4 minutes left to the hour and lasts for 2 minutes and 30 seconds (150 million years). Mammals and flowering plants dominate the landscape of the Cenozoic era during the last 1 minute and 15 seconds of the clock hour. Not until the closing seconds of the evolutionary hour does the human species appear on the scene. The whole span of recorded human history covers the last 3 seconds.

The Fossil Record

The general character of plant life from the dawn of the Cambrian period to recent times is depicted in figure 15.2. Only the broader aspects of past evolution are portrayed. It is evident that the flora has changed in composition over geologic time. New types of plants have continually appeared. Some types

Figure 15.1 Model of evolutionary time, where one second equals one million years. The Precambrian is represented as originating at 12 o'clock, or 3.6 billion years. The human species appeared about 3 million years ago, or a mere terminal three seconds of the evolutionary hour.

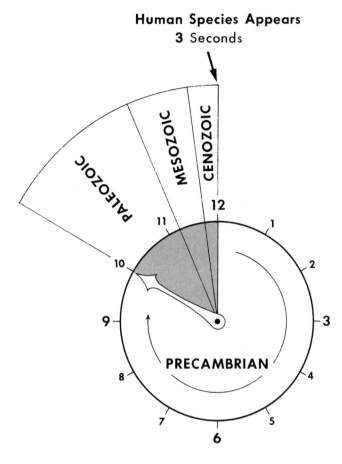

thrived for a certain time and then disappeared. Others arose and continued to flourish. Still others have survived until the present, only in much reduced numbers.

The Cambrian and Ordivician periods were characterized by a diverse group of algae in the warm oceans and inland seas. One of the most significant advances, which occurred during the Silurian period, was the transition from aquatic existence to life on land. The first plants that established themselves on land were diminutive herbaceous forms, the *psilophytes*, literally "naked plants," in allusion to their bare, leafless stems. The primitive woody stems

Figure 15.2 Historical record of plant life. Plant life during the Cambrian and Ordovician, the first two periods of the Paleozoic era, was confined to the water; seaweeds (algae) of immense size, often several hundred feet in length, dominated the seas. Land plants came into existence in Silurian time, in the form of strange little vascular plants, the psilophytes. In the Carboniferous period, imposing spore-bearing trees (Lepidodendrids and Calamites) and primitive naked-seeded plants (Pteriodosperms and Cordaites) reached their peak of development. The end of the Paleozoic era marked the extinction of the majority of the luxuriant trees of the Carboniferous coal swamps. The Mesozoic era was the *Age of Gymnosperms*, as evidenced by the abundance of cycads, ginkgoes, and conifers. Flowering plants (Angiosperms) rose to ascendancy toward the close of the Mesozoic era and established themselves as the dominant plant group on the earth.

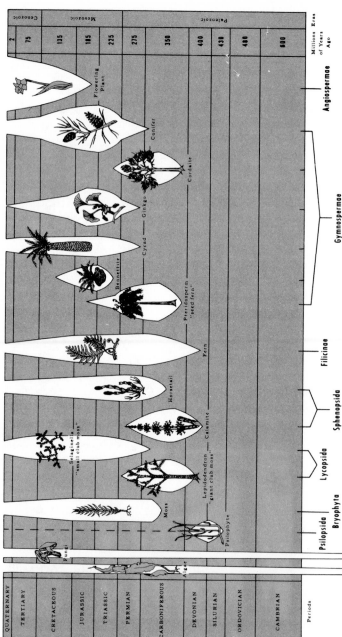

Figure 15.3 **Luxuriant forests of giant trees** with dense undergrowth flourished in the Great Coal Age (Carboniferous period), between 350 million and 275 million years ago. The massive trunks at the left are the Lepidodendrids, an extinct group whose modern relatives include the small, undistinguished club mosses *(Selaginella)* and the ground pines *(Lycopodium).* The tall, slender tree with whorled leaves at the right is a Calamite, represented today by a less prominent descendant, the horsetail *Equisetum.* The fernlike plants bearing seeds (at the left) are seed ferns (Pteridosperms), which resembled ferns but were actually the first true seed-bearing plants (Gymnosperms). (Field Museum of Natural History, Chicago.)

sent up swollen aerial shoots to scatter their spores and thrust branches down into the earth for water. The existence of psilophytes made possible the subsequent emergence of an infinite variety of tree-sized plants that flourished in the swamps of Carboniferous times. Two prominent trees that luxuriated in Carboniferous forests were the giant club moss, *Lepidodendron,* which soared to heights of 130 feet on slender trunks, and the coarse-leaved *Cordaites,* progenitors of the modern conifers. The first seed-bearing plants, *Pteridosperms,* arose in the humid environment of the vast swamps. These ancient groups of tall trees dwindled toward the close of the Paleozoic era and shortly became extinct. Their decomposed remains led to the formation of extensive coal beds throughout the world. Figure 15.3 shows the appearance of a Carboniferous swamp as reconstructed from fossils.

By far the greater number of Paleozoic species of plants failed to survive. The Paleozoic flora was largely replaced by the seed-forming gymnosperms, which became prominent in the early Mesozoic era. A diverse assemblage of cycads, ginkgoes (maidenhair trees), and conifers formed elaborate forests. During the closing years of the Mesozoic, the flowering plants (angiosperms), which began very modestly in the Jurassic period, underwent a phenomenal development and today constitute the dominant plants of the earth. The angiosperms have radiated into a variety of habitats, from sea level to mountain summits and from the humid tropics to the dry deserts. Associated with this diversity of habitat is great variety in general form and manner of growth. Many angiosperms have reverted to an aquatic existence. The familiar duckweed, which covers the surface of a pond, is a striking example.

In the animal kingdom, we witness a comparable picture of endless change, with waves of extinction and replacement (fig. 15.4). The deep Cambrian rocks contain the remains of varied invertebrate animals—sponges, jellyfishes, worms, shellfishes, starfishes, and crustaceans. One of the major developments of the Cambrian period was the advent of hard parts in the form of protective shells and plates. The shelled invertebrates were already so well developed that their differentiation must have taken place during the long period preceding the Cambrian. That this is actually the case has been attested by the discoveries of rich deposits of Precambrian fossils in rocks in the Ediacara Hills in southern Australia and on Victoria Island in the Canadian Arctic. These fossil-bearing rocks lie well below the oldest Cambrian strata, and contain imprints of jellyfishes, tracks of various kinds of worms, and traces of soft corals. Since the rocks on Victoria Island are 700 million years old, these discoveries extend the fossil record of invertebrates 100 million years beyond the Cambrian.

Dominating the scene in early Paleozoic waters were the crablike trilobites and the large scorpionlike eurypterids (fig. 15.4). Most trilobites were not more than 3 inches long; the so-called giant trilobite did not exceed 18 inches. The most powerful dynasts of the Silurian waters were the eurypterids, which occasionally attained a length of 12 feet. Common in all Paleozoic periods were the nautiloids, related to the modern *Nautilus*. The clamlike brachiopods, or lampshells, constitute nearly one-third of all Cambrian fossils, but are relatively inconspicuous today. The odd graptolites, colonial animals whose carbonaceous remains resemble pencil marks, attained the peak of their development in the Ordovician period and then faded away. No land animals are known for Cambrian and Ordovician times. Seascapes of the early Paleozoic are shown in figure 15.5.

The bony fishes arose in the Devonian, the Age of Fishes. The lobe-finned fishes, or crossopterygians, had strong muscular fins by which they crawled, in times of drought, along arid streambeds from pool to pool. These air-breathing crossopterygians gave rise to the first land-dwellers among vertebrates, the amphibians. The end of the Paleozoic era witnessed a decline in amphibians and the initiation of reptilian diversification.

Many of the prominent Paleozoic marine invertebrate groups became extinct or declined sharply in numbers before the Mesozoic era. During the Mesozoic, shelled ammonoids thrived in the seas, and insects became prominent. Some of the ancient insects were enormous, like *Meganeuron*, a form of dragonfly with a wingspread of 29 inches. Among vertebrates, the reptiles radiated swiftly to dominate the land. At the close of the Mesozoic, the once-successful marine ammonoids perished and the reptilian dynasty collapsed, giving way to the birds and mammals. Insects have continued to thrive and have differentiated into a staggering variety of species. Well over 800,000 different species of insects have been described, and conservative estimates place the total number of living species today at 3 million.

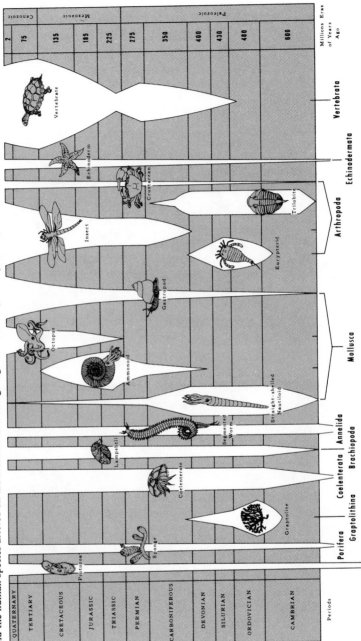

Figure 15.4 Historical record of animal life. Many of the invertebrate groups were already highly diversified and abundant in the Cambrian, the first period of the Paleozoic era, approximately 600 million years ago. The Paleozoic era is often called the *Age of Invertebrates*, with its multitude of nautiloids, eurypterids, and trilobites. Brachiopods with hinged valves were the commonest shellfish of the Paleozoic seas. In the Mesozoic era, air-breathing insects and vertebrates, notably the widely distributed reptiles, held the center of the stage. The Mesozoic seas were populated with large, shelled ammonoids, now extinct. Warm-blooded vertebrates (birds and mammals) became prominent in the Cenozoic era, and the human species arrived on the scene in the closing stages of this era (see chapter 16.)

204

Figure 15.5 Highly diversified assemblage of invertebrates of early Paleozoic seas. *Top:* Ordovician period. Large organism in foreground is a straight-shelled nautiloid. Other prominent forms are trilobites, massive corals, smaller nautiloids, and a snail. *Bottom:* Cambrian period. Conspicuous animals are the trilobite (center foreground), eurypterid (center background), and the jellyfish (left). Other animals include brachiopods, annelid worms, sea cucumber, and varied shelled forms. (*Top,* Field Museum of Natural History, Chicago; *bottom,* American Museum of Natural History.)

During the course of evolution, plant and animal groups interacted to each other's advantage. There is little doubt, for example, that the rise and spread of flowering plants fostered the diversification and dispersal of insects. As flowering plants became less and less dependent on wind for pollination, a great variety of insects emerged as specialists in transporting pollen. The colors and fragrances of flowers evolved as adaptations to attract insects.

Flowering plants also exerted a major influence on the evolution of birds and mammals. Birds, which feed on seeds, fruits, and buds, evolved rapidly in intimate association with the flowering plants. The emergence of herbivorous mammals coincided with the widespread distribution of nutritious grasses over the plains during the Cenozoic era. In turn, the herbivorous mammals furnished the setting for the evolution of carnivorous mammals. The interdependence between plants and animals continues to exist in nature today.

Convergence

The major pattern of the evolutionary panorama is divergence—the tendency of a population of organisms to become highly diversified when the population spreads over an area with varied habitats. The population can diverge into radically different lines, each modified for a specific ecological role. Now, essentially similar habitats may be found in widely separate parts of the world. It would thus not be surprising to find that two groups of organisms, unrelated by descent but living under similar environmental conditions in different geographic regions, can exhibit similarities in habits and general appearances. The tendency of one group of organisms to develop superficial resemblances to another group of different ancestry is called *convergence*.

Convergence is not an uncommon phenomenon in nature. Many unrelated, or remotely related, organisms have converged in appearance as a consequence of exploitation of habitats of similar ecological makeup. The relationship of convergence to adaptive radiation should be evident; the former is the inevitable result of the countless series of adaptive radiations that have taken place in scattered parts of the globe. A striking example of convergence is afforded by the living marsupials of Australia.

The massive inland continent of Australia has long been isolated from Asia, at least since the commencement of the Cenozoic era, the last 65 million years. It is on this isolated island that the marsupials—primitive pouched mammals—survived, free from the competition of the more efficient placental mammals when they came into prominence. In the absence of a land migration route between Asia and Australia, the latter land mass was inaccessible to practically all the placental mammals. On other continents, the marsupials perished, save for the peculiar American opossums, the didelphids.

Imprisoned in Australia, the marsupials spread into a variety of habitats. Several live in the open plains and grasslands; some are tree-dwellers; others are burrowers; and still others are gliders (fig. 15.6). Most kangaroos are terrestrial, but one variety, the monkeylike kangaroo, is arboreal. The slow-moving, nocturnal "teddy bear," or koala, lives and feeds on *Eucalyptus,* the dominant tree of Australia. The bandicoot, with rabbitlike ears, has sturdy claws adapted for digging in the ground in search of insects. Marsupial moles live in desert burrows, and flying phalangers have webs of loose skin stretched between the forelimbs and hindlimbs. The flying phalanger cannot actually fly, but is adept at gliding. The impressive diversity of marsupials thus represents an admirable example of adaptive radiation.

Figure 15.6 Marsupials, or pouched mammals, of Australia, illustrating the twin themes of adaptive radiation and convergence. The marsupials have radiated, or diversified, into a variety of forms, ranging from the tiny insect-eating "mouse" to the fierce, flesh-eating Tasmanian "wolf." Many of the marsupials resemble in appearance and habit the placental mammals of other parts of the world, although they are *not* closely related to them.

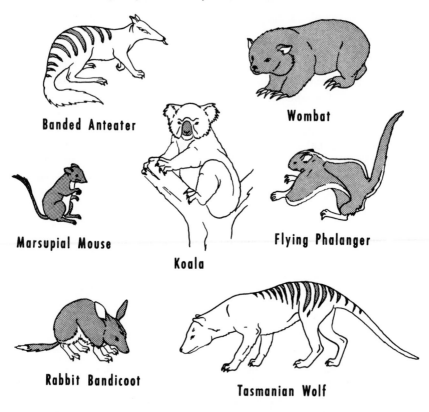

Banded Anteater

Wombat

Marsupial Mouse

Flying Phalanger

Koala

Rabbit Bandicoot

Tasmanian Wolf

The marsupials of Australia also vividly illustrate the phenomenon of convergence. They have filled the ecological niches normally occupied by placental mammals in other parts of the world. The marsupial "mouse," "mole," "anteater," "wolf," flying phalanger, and groundhoglike wombat strikingly resemble the true placental types—mouse, mole, anteater, wolf, flying squirrel, and groundhog, respectively—in general appearance and in ways of life.

It is interesting to note that a marsupial "bat" has not evolved in Australia. The opportunity was apparently denied by the invasion of placental bats from Asia, one of the few placental forms that managed, probably as a result of dispersal by flight and winds, to reach Australia. With the coming of humans, the secure existence of the marsupials has been threatened. Prehistoric humans introduced dogs, which ran wild (the dingos); later human settlers brought a number of European placental mammals, such as the rabbit, hare, fox, and Norway rat. Among the marsupials faced with extinction are the marsupial "wolf," which survives today only in Tasmania, and the slow-moving "anteater," which is rapidly disappearing.

The Plains-Dwelling Mammals

When the great majority of land mammals first arose, South America was just as much an island continent as Australia. An impressively rich assemblage of mammals evolved in isolation in South America during the 70 million years that the continent remained separated from the rest of the world. Such bizarre placental mammals emerged as the giant ground sloth, the enormous toothless anteater, and the armored armadillolike glyptodont. The predatory wolves that evolved in South America were not true carnivores; rather, they resembled the predaceous marsupials of Australia.

The spread of grasses across the flatlands of the Americas in the Tertiary period was associated with the rapid expansion of herbivorous hoofed mammals (ungulates) in both continents. The crowd of grazing mammals that arose is so bewildering that it is difficult to call up a true picture of the immense variety of ungulate types. These archaic hoofed mammals are quite unfamiliar to most people; they have no vernacular names and taxonomists have had to create special suborders for many of them. Selected examples of these curious hoofed mammals were shown earlier in figure 1.8 (chapter 1). These South American forms bear only superficial resemblances to the hoofed mammals in other parts of the world. *Toxodon* superficially resembles a rhinoceros, *Thoatherium* has the appearance of a horse, *Pyrotherium* may be likened to an elephant, and *Macrauchenia* looks like a camel. The similarities in external appearances signify convergence, and not relationship by descent.

At the time the Isthmus of Panama rose from the depths, the land bridge across the Bering Straits lay completely dry. Large mammoths streamed in increasing numbers across the Bering Strait from Asia into North America during the Pleistocene. The mammoths dominated the American plains, particularly the 10-foot-high woolly mammoth. North America was also the recipient of some of the South American mammals. Among the immigrants from south of the border were *Boreostracon,* a giant glyptodont, encased in armor and an offshoot of the armadillo line, and *Megatherium,* a great ground sloth that weighed more than a modern elephant (fig. 15.7).

The South American hoofed mammals, evolving independently of mammalian types in North America, thrived for many millions of years. Then the water gap between the Americas was closed. The Isthmus of Panama rose above the sea, establishing a land corridor linking the two continents. Through this land bridge infiltrated a horde of northern invaders. Predatory carnivores, such as saber-toothed tigers and wolves, entered South America from the north, as did progressive herbivorous ungulates, among which were tapirs,

Figure 15.7 **Prominent land mammals** that occupied the flatlands of the Americas in the Tertiary period.

Woolly Mammoth

Saber-toothed Tiger

Glyptodont

Ground Sloth

peccaries, llamas, and deer. All of the native South American hoofed mammals, without exception, were driven into extinction. South America also became deprived of its indigenous variants of ground sloths and glyptodonts. It then later lost several of the immigrants, notably the mastodons and saber-toothed tigers.

Evolutionary Stability

The multitude of different kinds of present-day organisms is impressive. Yet, the inhabitants of the world today are only a small percentage of the tremendous array of organisms of earlier periods. As we have seen, the fate of most lineages of organisms throughout time is extinction. Apparently, only those populations that can continue to adapt to changing environmental conditions avoid extinction. Yet, some types of organisms have not changed appreciably in untold millions of years. The opossum has survived almost unchanged since late Cretaceous, some 75 or more million years ago. The horseshoe crab, *Limulus,* is not very different from fossils uncovered some 500 million years ago. The maidenhair, or ginkgo, tree of the Chinese temple gardens differs little from its ancestors 200 million years back. The treasured ginkgo has probably existed on earth longer than any other present tree. Darwin called the ginkgo "a living fossil."

Long-standing stability of organization seems antithetical to the concept of evolution. Nevertheless, the notion that an established species may be rather stable is not completely at variance with conventional evolutionary thought. Long periods of relatively little change may be the outcome of stabilizing selection (see chapter 6), in which the typical members of a population are selectively favored relative to the extremes. In other words, the "status quo" is maintained by selection after an ecological niche in a habitat has been successfully filled. Admittedly, to some investigators, it seems implausible that natural selection could be so conservative as to continue to favor the same phenotype for many millions of years.

Selected Readings

Barghoorn, E. S. 1971. The oldest fossils. *Scientific American,* May, pp. 30–42.

Bock, W. J. 1972. Species interactions and macroevolution. *Evolutionary Biology* 5:1–24.

Colbert, E. H. 1955. *Evolution of the vertebrates.* New York: John C. Wiley & Sons.

Eglinton, G., and Calvin, M. 1967. Chemical fossils. *Scientific American,* January, pp. 32–43.

Fingerman, M. 1975. *Animal diversity.* New York: Holt, Rinehart & Winston.

Glaessner, M. F. 1961. Precambrian animals. *Scientific American,* March, pp. 72–78.

Hanson, E. 1964. *Animal diversity.* Englewood Cliffs, N.J.: Prentice-Hall.

Lewin, R. 1982. *Thread of Life*. Washington, D.C.: Smithsonian Books.

Lewontin, R. C. 1978. Adaptation. *Scientific American,* September, pp. 212–30.

May, R. M. 1978. The evolution of ecological systems. *Scientific American,* September, pp. 160–75.

Raup, D. M., and Stanley, S. M. 1978. *Principles of paleontology.* San Francisco: W. H. Freeman and Co.

Romer, A. S. 1945. *Vertebrate paleontology.* Chicago: University of Chicago Press.

Rush, J. H. 1963. *The dawn of life.* New York: The New American Library, Signet Library.

Schopf, J. W. 1978. The evolution of the earliest cells. *Scientific American,* pp. 110–138.

Simpson, G. G. 1950. History of the fauna of Latin America. *American Scientist* 38:361–89.

Stirton, R. A. 1959. *Time, life, and man.* New York: John C. Wiley & Sons.

Valentine, J. W. 1978. The evolution of multicellular plants and animals. *Scientific America,* September, pp. 140–58.

Walton, J. 1953. *An introduction to the study of fossil plants.* London: A. & C. Black.

Windley, B. F. 1977. *The evolving continents.* New York: John C. Wiley & Sons.

16
Emergence of the Human Species

The human species is the outcome of the same natural evolutionary forces that have shaped all animal life. Humans are a recent arrival on the earth, representing the latest adaptive advance in the animal kingdom. Although we occupy a unique position in the living world, we are clearly part of the fabric of nature and have ancestral ties with other animals. There is almost universal unanimity that our closest relatives are the apes. The line leading ultimately to humans diverged from the ape branch during Tertiary times. There are several candidates among Tertiary fossil apes that qualify as ancestors of the human species. One often hears stated, hesitatingly and perhaps in the form of an apology, that we are not really closely related to the apes but merely share a very distant ancestry with the apes. By indirection, the evasive idea is conveyed that our remote generalized ancestor would not be at all apelike. This is sheer deception. There is simply no escaping the idea that our very early predecessors had many features in common with the apes.

Primate Radiation

Primates, the order to which humans belong, underwent adaptive radiation in Cenozoic times, approximately 65 million years ago. The primates are primarily tree dwellers; only humans are fully adapted for life on the ground. Many of the noteworthy characteristics of the primates evolved as specializations for an arboreal mode of life. Depth perception is important to a tree-living animal; the majority of the primates are unique in possessing binocular, or steroscopic, vision, wherein the visual fields of the two eyes overlap. The hands evolved as prehensile organs for grasping objects, an adaptation later useful to humans for manipulating tools. The use of the hands to bring objects to the nose (for smelling) and to the mouth (for tasting) was associated with a reduction of the snout, or muzzle, and a reduction of the olfactory area of the brain. Primates generally have a poor sense of smell. Closely associated with enhanced visual acuity and increased dexterity of the hands was the marked expansion of the brain, particularly the visual and memory areas of the brain. Progressive enlargement of the brain culminated, in humans, in the development of higher mental faculties.

The primate stock arose and differentiated from small, chisel-toothed, insectivorous ancestors (fig. 16.1). The Asiatic tree shrew, *Tupaia,* an agile tree climber that feeds on insects and fruits in tropical forests, survives as a model for the ancestor of the primates. Most authorities place the tree shrew in the mammalian order. Insectivora, while a few hold the view that the tree shrew is a primitive, generalized member of the primates. The order Primates is generally divided into two groups or suborders, the *Prosimii* and the *Anthropoidea.* The prosimians are arboreal and largely nocturnal predators; they include

Figure 16.1 **Adaptive radiation of the primates** from a basic stock of small, insect-eating placental mammals, the Insectivora (whose living kin include the shrews, moles, and hedgehogs). Based on paleontological and anatomical evidence, the pongid (ape) and hominid (human) lineages diverged during the Miocene epoch, about 20 million years ago. Estimates based on biochemical data are presently at variance with the fossil date. The biochemical evidence suggests that the lineage leading to the human split off from the ape line only about 5 million years ago.

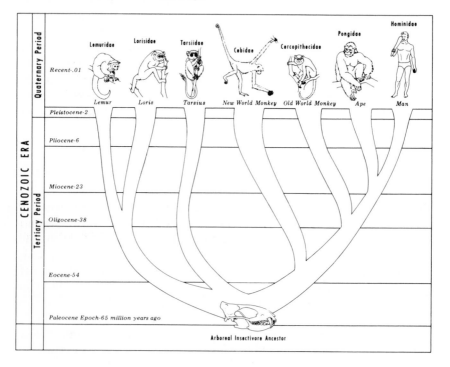

such tropical forms as the lemurs, the lorises, and the tarsiers. As adaptations to an arboreal-nocturnal niche, these prosimians evolved large, forwardly placed eyes and strong, grasping hands and feet. The lemurs are largely confined to Madagascar, the lorises are found principally in eastern Asia, and the tarsiers are limited to southeastern Asia. The tarsiers represent the most advanced group of prosimians, and may be said to foreshadow the higher primate trends of the Anthropoidea.

The more advanced primates, the anthropoids, are composed of the New World monkeys, the Old World monkeys, the apes, and humans. All are able to sit in an upright position, and thus the hands are free to investigate and manipulate the environment. The monkeys are normally quadrupedal, running along branches of trees on all fours. Among the apes, gibbons habitually use their arms for hand-over-hand swinging in a motion known as brachiation. The great apes (orangutan, chimpanzee, and gorilla) can maintain prolonged semierect postures. When on the ground, the chimpanzee and gorilla are "knuckle-walkers," the weight being placed on the knuckles of the hands rather than on the extended fingers. The chimpanzee and the gorilla regularly travel quadrupedally on the ground. Humans alone are specialized for erect bipedal locomotion.

The anatomical and physiological features that distinguish modern humans from the living apes are comparatively few. The resemblances in skeletal structures, muscular anatomy, physiological processes, serological reactions, and chromosome patterns are all strikingly close. Recent biochemical studies emphasize that the two African apes, chimpanzee and gorilla, bear a special evolutionary relationship to humans. As discussed in chapter 14, fibrinopeptide molecules are involved in the process of blood clotting and consist of short chains of amino acids. The fibrinopeptide molecules in different mammalian species have undergone rapid and numerous changes in amino acid sequences. It is impressive that all the amino acid residues of this molecule are the same and appear in identical sequences in the chimpanzee, gorilla, and humans. The differences in amino acid sequences of fibrinopeptides have been found to be as follows: from human to chimpanzee, 0; to gorilla, 0; to orangutan, 2; to gibbon, 5; to Old World monkeys (green monkey and Rhesus macaque), 5; to New World monkeys (spider monkey and cebus monkey), 9–10; and to a prosimian (slow loris), 18. Humans clearly are more closely related to the African apes than either is to the Asian apes (the orangutans).

Studies of the DNA molecule have also disclosed the extraordinary close relationship between the African apes and humans. The complementary strands of a double-stranded DNA molecule can be separated experimentally into two single threads. A dissociated strand from one primate can then be rejoined to a single strand from another primate. The degree of reassociation of the two different strands into a hybrid double-stranded DNA molecule is dependent on temperature. A hybrid double-stranded product is less stable than the pure DNA molecule, with its nearly perfect matching of nitrogenous bases. The less precise the matching of bases, the lower the thermal stability of the double-stranded DNA molecule. Stated another way, the greater the thermal stability of the hybrid DNA, the greater the degree of evolutionary relationship. Not surprisingly, the DNA of humans chemically agrees most closely with the DNA of a chimpanzee. Only relatively few nucleotide changes occurred between the two groups since the time of their divergence. The biochemical evidence indicates that there could be only one ancestor for humans and the African apes. Indeed, based on the molecular data, the lineage leading to humans split off from the ape line a mere 5 million years ago.

The pronounced differences between humans and apes relate mainly to locomotory habits and brain growth. Humans have a fully upright posture and gait and an enlarged brain. The cranial capacity of a modern ape rarely exceeds 600 cubic centimeters, while the average human cranial capacity is 1,350 cubic centimeters. The mastery of varied environments by humans has largely been the result of superior intelligence, gradually acquired throughout evolution.

It may be difficult to envision humans other than as they exist today. Nevertheless, it should not be imagined that fossil specimens of the early Pleistocene epoch possessed the attributes of contemporary forms. Present-day humans are certainly different from their predecessors, in much the same manner that modern horses are different from their forerunners. It is important to recognize that there have been different kinds of human beings. The evolutionary process of adaptive radiation led to a family of humans, recognized formally as the Hominidae, of which our modern species—*Homo sapiens*—is the sole survivor.

Forerunners of the Great Apes

The modern apes are but a remnant of a once diverse and widespread group. The great apes reached their evolutionary peak in the African forests of the early Miocene, some 20 million years ago. The most famous of the Miocene apes has been classified as *Dryopithecus,* the oak-ape, so called because of the presence of oak leaves in the fossil beds. Known in more informal terms as the dryopithecines, these primitive oak-apes were a relatively successful group, having flourished for at least 10 million years. At that time, the Arabian peninsula joined Africa and Eurasia together, and the dryopithecines ranged widely throughout the two great land masses. The dryopithecines are clearly members of the ape family (Pongidae).

The African variety, *Dryopithecus africanus*—first named *Proconsul,* after a famous chimpanzee at the London Zoo—may not be far removed from the common stock from which apes and humans arose. Although primarily a tree dweller, *Dryopithecus africanus* apparently wandered on the ground and profited from the vegetable foods in the open grasslands (fig. 16.2). The transition from tree dwelling to ground living might well have first appeared at this time.

Figure 16.2 **Reconstruction** of *Dryopithecus africanus,* an apelike type that prowled East Africa 20 to 25 million years ago. This pongid apparently led an agile life both on and off the ground. (Painting by Maurice Wilson; by permission of the Trustees of the British Museum—Natural History.)

The Miocene epoch was characterized by expanding populations of varied species of monkeys and apes in the African forests. Population pressure probably contributed to the exploratory departure of oak-apes from the forest fringes to the savanna grasslands. These pioneer ground-dwelling apes apparently were able to establish themselves in the open bushy plains by subsisting mainly on plant foods and largely avoiding confrontations with the savanna carnivores. But as the forests continued to recede and yield to the grasslands, the dryopithecines became extinct. Some species of this ape may have evolved into the still surviving orangutan, gorilla, and chimpanzee. Even these modern apes live in steadily shrinking forest habitats and are likely to face extinction in the wild by the end of this century.

Forerunners of the Hominidae

Included in the great Miocene diversity of apes in Africa and Eurasia were two enigmatic apes with hominid-like dentition, called *Ramapithecus* and *Sivapithecus.* Both were named after dieties—Rama, a Hindu god, and Siva, a god of destruction. Ever since the first fragile upper jaw of a ramapithecine was found by G. Edward Lewis in the early 1930s in the Silawik Hills of northern India, anthropologists have argued over whether these fossils were the first ancestral hominid, an ancestral pongid (in particular, a proto-orang-utan), or merely an evolutionary dead end.

The striking ramapithecine fossil jaws exhibit reduced canine teeth, relatively small incisors, and low-crowned molars capped with a thick layer of enamel. The thickly enameled molar teeth suggest that *Ramapithecus* ate chiefly hard-covered fruits, nuts, and seeds rather than the softer forest fruits usually consumed by apes. Additionally, the flat wear of the molars hints of a grinding type of mastication. Such grinding action would be unusual for apes, who tear and crush, rather than masticate, their succulent food.

When the jaw of *Ramapithecus* was originally reconstructed, the dental arcade was presumed to be parabolic in shape, as in hominids (fig. 16.3). The smoothly rounded contours of the arcade suggested that *Ramapithecus* had, in incipient form, the dental and facial organization characteristic of later hominids. In the 1960s, Elwyn Simons of Yale University championed the view that *Ramapithecus* may be close to the base of the stem of the hominid family. That is to say, this fossil form marks the point where the hominid lineage separated from the pongid assemblage. In fact, it became an acceptable view that *Ramapithecus* was the earliest member of the human family (Hominidae).

In the last few years, there has been a marked inclination to dethrone *Ramapithecus* as a hominid. With the discovery of complete jaws of *Ramapithecus,* the dental arcade is unmistakably V-shaped, rather than parabolic (fig. 16.3). Adrienne Zihlman of the University of California, Santa Cruz, assesses *Ramapithecus* as merely one of several kinds of apes that ventured into open country more than 14 million years ago. The recent discovery of fossilized limb bones from Pakistan suggests that *Ramapithecus* was more quadrapedal than previously thought, and more aboreal than terrestrial. The Harvard anthropologist David Pilbeam favors the view that *Ramapithecus,* along with *Sivapithecus,* represent groups that were ancestral to the orang-utan, which split off from the lineage leading to hominids about 16 million years ago.

If the apelike *Ramapithecus* is on the line to the common ancestry of both apes and humans, it is probably distantly removed from the actual pongid-hominid divergence. As previously mentioned, data from molecular studies suggest that the bifurcation between pongids and hominids occurred no more than 5 million years ago. Investigators who persist in advocating the hominid status of *Ramapithecus* dismiss the molecular evidence by claiming that the molecular clock is not finely tuned and keeps inaccurate time. The resolution of the actual time of the pongid-hominid split awaits the unearthing and analysis of further fossil remains dating between 4 and 16 million years ago.

Figure 16.3 **Comparisons of dental arcades** of chimpanzee *(left),* human *(right),* and *Ramapithecus (middle).* In a once-accepted reconstruction of the jaw of *Ramapithecus,* the shape was conceived as parabolic, like that of humans. Recent evidence indicates that the dental arcade of *Ramapithecus* is V-shaped. Note that the long, projecting canines of the chimpanzee contrast with the relatively shortened human canines.

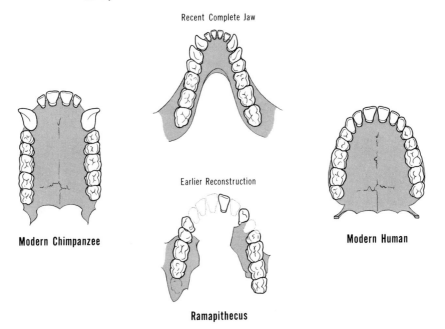

Recent Complete Jaw

Earlier Reconstruction

Modern Chimpanzee

Modern Human

Ramapithecus

The First Hominids

There is, at present, an inexplicable 4-million-year gap between the disappearance of *Ramapithecus* (about 8 million years ago) and the arrival of the first undeniable hominid—*Australopithecus*—about 4 million years ago. The australopithecine stage is a long phase of slow evolution in which several species of early hominids existed. The initial stage commenced four million years ago and ended about one million years ago.

In 1924, the Australian anatomist, Raymond A. Dart, announced the discovery of an unusual small skull from a Pleistocene limestone quarry near the village of Taung in the Transvaal region of South Africa. The fossilized skull was that of a child of about six years. The little Taung skull bears some resemblance to the skull of a young chimpanzee, but many of its components, notably the teeth, show pronounced affinities to humans. Certain striking features of the Taung skull suggest that the child had walked upright. The remarkable skull was designated by the formidable name of *Australopithecus africanus* (*austral,* for "south," and *pithekos,* for "ape"). Dart was confident that *Australopithecus* was related to the ancestral stock of humans, rather than to the great apes. In other words, the Taung fossil represented an early member of the Hominidae, the family of humans, rather than of the Pongidae, the family of apes.

Dart's declaration, which several scientists initially derided, was fortified by findings in the 1930s by the late Robert Broom, a Scottish paleontologist. Adult skulls of *Australopithecus* were dug out from caves in Sterkfontein, Kromdraai, and Swartkrans in South Africa. The adult skulls confirmed the hominid anatomical pattern seen in the juvenile Taung cranium. The several new fossil forms were originally given different names. However, in recent years, it has become customary to refer to the South African fossils collectively as the australopithecines. They were short, four to five feet in height, with a small ape-sized brain (cranial capacity range of 450 to 600 cubic centimeters). Nonetheless, the australopithecines stood upright, walked bipedally, and dwelt in open country (fig. 16.4). These circumstances nullify the popular view that humans were intelligent animals when they first came down out of the trees. It seems clear that erect bipedal locomotion on the ground preceded the development of a large complex brain.

The australopithecines are decidedly early representatives of the hominid lineage. However, spirited debate exists over which of the australopithecines occupy a prominent place in the direct ancestry of humans. Current evidence favors the idea that there were two distinct species of australopithecines—the light-jawed, slender ("gracile") *Australopithecus africanus* and the heavy-jawed, robust *Australopithecus robustus* with extremely large grinding molars (fig. 16.5). The former species was not over four feet tall and weighed no more than 60 pounds, whereas the latter species was a foot taller and at least 30 pounds heavier. When two contemporaneous species occupy the same habitat, ecologists have observed that the potential competitors become differentially specialized to exploit different components of the local environment. Thus, direct competition for food resources is minimized and the two species are able to coexist. It is thought that *Australopithecus robustus* became or remained exclusively vegetarian, whereas *Australopithecus africanus* increasingly supplemented its diet with animal food. The dietary difference is supported by the finding that *A. africanus* had smaller molars than *A. robustus*.

If, as it appears likely, the two *Australopithecus* species did coexist at the same time in the same region, then only one of the two could have been the progenitor of a more modern species of man. It has been suggested that the vegetarian *A. robustus* perished without leaving any descendants, and that *A. africanus* was the forebear of a more advanced hominid. The dietary change to carnivorism may have represented one of the most important steps in transforming a bipedal ape into a tool-making and tool-using human.

The South African finds have been supplemented and extended by fossil discoveries by the late Louis Leakey and his wife Mary at the 25-mile-long Olduvai Gorge in Tanzania. Olduvai Gorge is situated in a volcanically active region in eastern Africa, and has invaluable sediments of lava from prehistoric volcanic activity. Since lava contains argon 40, the daughter isotope of potassium 40, the rate of decomposition from potassium to argon permits an accurate dating of the Olduvai layers. Like a miniature Grand Canyon, the sides of Olduvai Gorge display different strata laid bare by the cutting of an ancient river. In an exposed stratum of the gorge, the Leakeys in 1959 uncovered bony fragments of a robust australopithecine, characterized by extremely massive jaws. This heavy-jawed fossil form with enormous teeth was called *Zinjanthropus,* or the Nutcracker Man. Fossil remains of *Zinjanthropus* were found in strata judged, by the potassium-argon dating method (instead of the conventional uranium-lead technique), to be about 1.8 million years old. Most scientists today agree that *Zinjanthropus* is essentially an eastern African variety of *Australopithecus robustus.* Others contend that *Zinjanthropus* has

Figure 16.5 Phylogeny of humans. *Australopithecus afarensis* followed by *Homo habilis* and *Homo erectus* are believed to have existed successively in geological time and to form a lineage leading to *Homo sapiens*. In the distant past, the *Homo* assemblage co-existed with the *Australopithecus* group. (Based on views by Donald Johanson.)

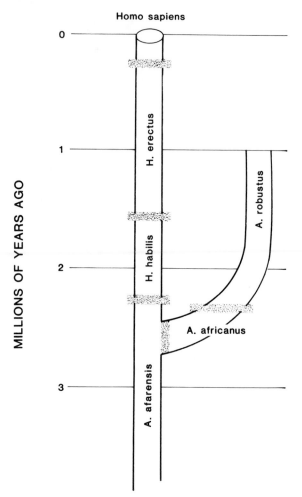

more exaggerated, or coarser, features than *A. robustus*, and warrants recognition as a separate species, *Australopithecus boisei*. It would thus represent a third species in the australopithecine complex.

Early in 1960, Louis and Mary Leakey found remains of a light-jawed hominid that they claimed was more advanced and more humanlike than *Australopithecus africanus*. Estimates of the brain capacities of the skulls averaged 637 cubic centimeters. This light-jawed type of Olduvai Gorge was said to be the first civilized or humanized being, deserving of the rank of *Homo*, namely *Homo habilis* (fig. 16.5). The specific name *habilis* means "handy," from the inferred ability of this hominid to make stone tools. The recognition of *Homo habilis* indicates that this primitive human being was evolving alongside the less hominized australopithecines, and lived side by side with them. The coexistence of *Homo habilis* and the australopithecines has been the subject of lively debate. Some writers argue that *Homo habilis* did not cross the threshold into the human grade, and represented only an advanced member of *Australopithecus africanus*.

Spirited controversy surrounds the discovery by the son, Richard Leakey, in 1972 of an unusual skull from the dessicated fossil beds of Lake Turkana in Kenya. This nearly complete skull was initially placed at 2.8 million years old, but is currently acknowledged as being only 1.8 million years old. The specimen has been cautiously designated only by its museum identification number—"1470." The cranial capacity of the "1470" skull measures 780 cubic centimeters, which is significantly larger than any australopithecine specimen. Skull "1470" clearly belongs to the genus *Homo,* although no species name has yet been designated. It most likely represents *Homo habilis,* and may be the linear ancestor of *Homo erectus* (fig. 16.5).

In 1975, Mary Leakey and her co-workers discovered fossil jaws and teeth in sediments at a flat expanse known as Laetolil, 25 miles south of Olduvai Gorge. These hominid jaw fragments bear a strong resemblance to the mandible and teeth of skull "1470." The Laetolil specimens are impressive in that they have been dated at 3.6 to 3.8 million years. Mary Leakey has classified the jaw fragments as the remains of the eastern African *Homo habilis.* If this is the case, the existence of *Homo habilis* has been extended back to at least 3.5 million years ago; accordingly, this human species is older than *Australopithecus africanus*. This would indicate that *Homo habilis* is not the direct descendant of the australopithecines.

One might suppose that an older, as yet undiscovered australopithecine (from Pliocene sediments) gave rise to *Homo habilis.* The candidate for such an ancestor appears to be the fossil hominid recovered in 1973 by Donald Johanson, curator at the Cleveland Museum of Natural History. The remains of this hominid were found at a formation in the Hadar region of Ethiopia, which has been dated between 2.6 and 3.3 million years ago. The most complete adult skeleton uncovered has been called "Lucy." She was only 3.5 to 4.0 feet tall and died when she was in her early twenties. Johanson contends that the hominid remains from Hadar and Mary Leakey's Laetolil fossils are so similar morphologically as to belong to a single human lineage. Both the Hadar and Laetolil specimens have been assigned to the same species and delineated as a new species, *Australopithecus afarensis* (fig. 16.5). The new species derives its name from the Afar locality in Hadar, Ethiopia, which yielded the most numerous specimens. Some anthropologists favor placing the Hadar and Laetolil specimens in the genus *Homo.* At present, it is safe to state that the *afarensis* remains, dated between 3 and 4 million years ago, constitute the earliest definitive members of the family Hominidae.

The interpretation of hominid phylogeny presented in figure 16.5 relies heavily on the notion that *Australopithecus afarensis* provides a suitable candidate for a form that was ancestral to the two lineages that existed contemporaneously, *Homo* and the australopithecines. The phylogenetic relationships, however, are widely debated, as may be deduced from the alternative schemes depicted in figure 16.6. Some anthropologists have held that *Australopithecus africanus* was directly ancestral to *Homo habilis;* others are convinced that *Homo habilis* co-existed with the gracile *Australopithecus africanus.*

Gathering-Hunting Way of Life

The lower Pleistocene hominids *(Homo habilis)* fabricated crude tools out of pebbles. The ability to fashion and wield tools followed, or accompanied, the emancipation of the hands that bipedal locomotion made possible. The open African savanna offered to the early plains-dwelling hominids a wider selection of foods than were available in the tropical forests. Tools were used to gather a variety of plant materials (nuts, seeds, fruits, tubers, and roots) and small animals (lizards and rodents). It is unlikely, as popularly imagined, that

Figure 16.6 Alternative phylogenic trees of the hominids. Although not encompassing all proposed schemes, these four are representative of currently held hypotheses. All concur in suggesting that *Australopithecus afarensis* is not far removed from the base of the stem of the hominid family.

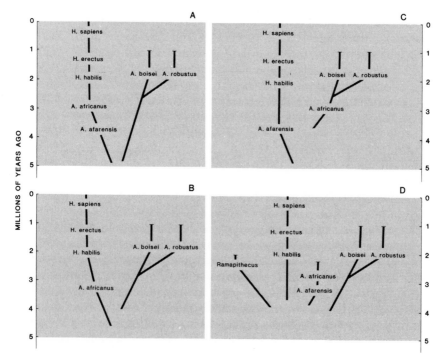

tools were first contrived as efficient spears to hunt large game or that our early ancestors were voracious meat-eating, predatory primates. The human species became accomplished hunters of the large animals only later in evolutionary history. The early hominids on the savanna were most likely primarily gatherers of plants and small animal life, and both males and females were opportunistic in devising tools for digging and processing plant foods. The gathering of food was particularly critical to females with dependent young. There was probably some stalking and killing of game, predominantly engaged in by males, but the uncertainty of success made hunting secondary or supplementary to the collecting of plant food with its assurance of success. The exploitation of food sources by both males and females on the savanna fostered the collaborative interactions of individuals. Cooperative behavior promoted the sharing of food for the first time. The sharing of food regularly is a social achievement unique to humans; only rarely do apes share food. The exchange of food is considered to be the earliest expression of human social reciprocity.

Homo Erectus

The earliest and best-known representative of *Homo erectus* is the famous Java fossil, first described as *Pithecanthropus erectus*. This primitive human being was discovered at Trinil, Java, in 1894 by Eugene Dubois, a young Dutch army surgeon. Dubois had been profoundly influenced by the writings of Charles Darwin, and had become imbued with the idea that he could find the origins of humans in the Far East. He surprised the world with the discovery of this early human. Curiously, Dubois, in his later years, inexplicably doubted his own finding and contended that *Pithecanthropus erectus* was merely a giant manlike ape. In the 1930s, additional fossil finds of *Pithecanthropus* were unearthed in central Java by the Dutch geologist G. H. R. von Koenigswald. The new findings confirmed the human status of the pithecanthropines.

The pithecanthropines lived during middle Pleistocene times, between 1.3 million and 300,000 years ago. These low-browed hominids were toolmakers and hunters who had learned to control fire. They probably had some powers of speech. Their ability to exploit the environment is reflected in the expanded size of the brain. The cranial capacity of the pithecanthropines was in the range of 775 to 1,000 cubic centimeters.

In the 1920s, elaborate excavations undertaken by the Canadian anatomist Davidson Black of caves in the limestone hills near Peking, China, led to the discovery of another primitive human, *Sinanthropus pekinensis,* or Peking (fig. 16.7). The cranial capacity in the Sinanthropines varied from 900 to 1,200 cubic centimeters. They fashioned tools and weapons of stone and bone, and kindled fire. There is a strong suspicion that *Sinanthropus* was cannibalistic and savored human brains, for many of the fossil braincases show signs of having been cracked open from below.

The Java and Peking hominids were originally each christened with a distinctive Latin name, *Pithecanthropus erectus* and *Sinanthropus pekinensis,* respectively. There is, however, no justification for recognizing more than the single genus of humans, *Homo.* Accordingly, modern taxonomists have properly assigned both the Java and Peking fossils to the genus *Homo.* Moreover, the morphological differences between these two fossil humans are readily within the range of variation that we observe in living populations today. These forms thus represent two closely related geographic races (subspecies) of the same species. Lastly, both Java and Peking hominids are distinct enough from modern humans *(Homo sapiens)* to warrant being placed together in different species, *Homo erectus* (fig. 16.5).

These nomenclatural changes may appear to be trivial, but the implications are great. One important implication is that human populations at any one time level were differentiated into geographical races of one species—not into distinct species or even genera. We can envision, for example, that approximately 500,000 years ago, there existed a single widespread species of humans, with eastern populations represented by the Java and the Peking variants (*Homo erectus erectus* and *Homo erectus pekinensis,* respectively) and the western populations constituted by types resembling Heidelberg (found in an early Pleistocene sand deposit near Heidelberg, Germany). The suspected wide distribution of *Homo erectus* has been confirmed by the recent discoveries of this early type of human in northern Africa (Ternifine, Algeria and Salé, Morocco) and eastern Africa (Olduvai Gorge and Koobi Fora). Evidently then, populations of *Homo erectus* spread successfully through the continents, from the tropical regions of Africa to southeast Asia. The widely dispersed *Homo erectus* adapted to a variety of changing climates and environments. As *Homo erectus* wandered from place to place, the hunting of game most likely became intensified. Vegetable foods continued to be important in subsistence, but the hunting habit was the way of life for nomadic *Homo erectus*. There is no evidence that the female participated in the hunting of large game; the adult female was increasingly encumbered with a fetus or by the care of the young, or by both. The long period of dependency of the young strengthened mother-child bonds, but also restricted the mobility and activity of the woman. It appears likely the female remained at the home base as a food gatherer while the male alone engaged in hunting. The immobility of the female and the prolonged immaturity of the young, coupled with the limitations imposed on the male in the number of females he could possibly support, apparently transformed a basically polygamous society into a monogamous structure.

A primary human innovation was relatively permanent pair-bonding, or monogamy. Sustained pair-bonding proved to be advantageous in several respects. It served to reduce sexual competition among the males. The prolonged male-female pairing increased the probability of leaving descendants. The heterosexual pair-bonding relationship became fortified as the estrus cycle of the female became modified into a condition of continuous sexual receptivity. The sustained sex interests of the partners made possible by the obliteration of estrus in the female increased the stability of the family unit, and facilitated the development of permanent family-sized shelters for rest, protection, and play. In essence, strong interpersonal bonds between a male, a female, and their children became the basis of the uniquely human family organization. The human male had become incorporated in the mother-young social group of the monkeys and apes.

Human language was fostered as males and females recorded experiences with each other and transmitted information to their children. Speech favored cooperation between local groups and the fusion of small groups into larger communities. Speech also fostered the successful occupation of one geographical area after another. The exchange of ideas over wide areas permitted human cultures of great complexity to develop.

Emergence of Modern Humans

One might expect that the fossil record of the last 200,000 years would present an unmistakably well-defined picture of the evolution of our modern species, *Homo sapiens*. This is not the case. The final stages leading to contemporary humans present several unresolved problems. We have yet to disentangle the transitional steps from the *Homo erectus* level of humans to the *Homo sapiens* grade. Moreover, the grade of *sapiens* contains at least two contrastingly different anatomical types—the bulkily built and heavily muscled Neanderthals and the slim-bodied Cro-Magnons.

The classic Neanderthal was first unearthed in 1856 in a limestone cave in the Neander ravine near Düsseldorf, Germany (fig. 16.8). It is one of the best known of fossil hominids, having been subsequently found at numerous widely separate sites in Europe, particularly in France. The Neanderthals were cave dwellers, short (about 5 feet) but powerfully built, with prominent facial brow ridges. They had large brains with an average capacity of 1,450 cubic centimeters, as opposed to 1,350 cubic centimeters in modern humans.

The Neanderthals were once popularly portrayed as brutish and dull-witted human creatures. On the contrary, the Neanderthals made complex stone tools, were accomplished hunters of large game, and withstood the rigors of the bitter cold climate of the last glaciation. There is evidence that the Neanderthals buried their dead with various ritual objects. The Neanderthals roamed over Europe up to about 40,000 years ago (Upper Pleistocene), and then dramatically disappeared. They were replaced by humans of a modern type, much like ourselves, which have been grouped under the common name of Cro-Magnon. Were the Neanderthals our immediate ancestors, or were they sterile

Figure 16.8 Neanderthal, a rugged cave-dweller who roamed Europe and the Middle East about 75,000 years ago. (Painting by Maurice Wilson; by permission of the Trustees of the British Museum—Natural History.)

offshoots of the human family tree? Those anthropologists who regard the Neanderthals as an evolutionary dead-end surmise that modern humans were evolving elsewhere on earth during the Neanderthal era in Europe.

The transition from the Neanderthal grade to the Cro-Magnon level is problematical. Prior to the time of the Neanderthals themselves, about 200,000 to 300,000 years ago, there emerged kinds of hominids, such as Swanscombe from England (fig. 16.9) and Steinheim from Germany, that did not conform to the classic Neanderthal type. The modern-looking yet ancient Swanscombe and Steinheim skulls are morphologically intermediate between *Homo erectus* and the later European Neanderthals. These specimens are apparently samples of a very early Neanderthal. The picture is far from clear. In the Near and Middle East, about 70,000 to 40,000 years ago, there existed an exceptionally heterogeneous group of humans. These are represented by specimens dug out of caves (Tabūn and Skhūl) on the slopes of Mount Carmel (in what is now Israel) and excavated from a cave at Shanidar in the mountains of northern Iraq. These fossil hominids ranged from individuals with almost typically western European Neanderthal features to individuals that are barely distinguishable from modern humans.

Figure 16.9 Reconstruction of the life of Swanscombe from England, a hominid morphologically close to both *Homo erectus* and the Neanderthals. The Swanscombe specimens are most likely very early *Homo sapiens,* dating back about 250,000 years ago. (Painting by Maurice Wilson; by permission of the Trustees of the British Museum—Natural History.)

Another element of uncertainty is the relation of the European Neanderthals to other Upper Pleistocene types in widely scattered parts of the world—Rhodesian of central Africa and the Solo of central Java. The Rhodesian specimens have a heavy skull and receding forehead like the European Neanderthals, but their limb bones differ in being more slender. Most authorities have considered the Rhodesian and Solo types to be the geographical equivalents of the Neanderthals of Europe. The Solo type has been characterized as a "tropical Neanderthal." This would lead us to believe that the classic Neanderthals were widely distributed in the Old World. These findings present knotty problems.

Status of the Neanderthals

Authorities do not all agree on the status of the Neanderthals. The populations of Neanderthals in Europe disappeared rather abruptly and were replaced by a modern group (Cro-Magnon) that definitely belongs to *Homo sapiens.* Does this indicate that two separate species, *Homo neanderthalensis* and *Homo sapiens,* actually existed together in Europe, and that the latter species displaced the former? Or were the Neanderthals ancestral to our modern species?

Some authorities assert that the classic Neanderthals were not directly involved in the ancestry of modern humans. Clark Howell, professor of anthropology at the University of Chicago, suggests that little, if any, opportunity existed for the exchange of genes between the Neanderthal population of western Europe and populations in the Middle East or elsewhere. During the Pleistocene epoch, glacial ice sheets covered many parts of Europe. It may be that the Neanderthals of western Europe were geographically isolated by the Ice Age of 100,000 years ago. The Neanderthals may have perished in isolation before modern humans arrived, or may have been overrun by more progressive Middle East contemporaries who had spread into western Europe. Under this viewpoint, the Neanderthals represented an evolutionary dead end.

The disagreements on the status of the Neanderthals continue to persist. In recent years, Erik Trinkaus and William W. Howells, anthropologists at Harvard University, have contended that the Neanderthals are in the direct line of ancestry to contemporary humans. Not all Neanderthals exhibited the extreme traits of the classic European type. The Neanderthals were actually widespread in distribution and were highly variable as a group. The varied Neanderthals belonged to the gene pool of *Homo sapiens.* The assemblage of *Homo sapiens* includes the early varieties (the Swanscombe-Steinheim stock), the later varieties of Neanderthals, and finally the modern type, the Cro-Magnons. Under this view, the Neanderthals are treated as different from modern *sapiens* only at the subspecies level. Thus, the Neanderthals are to be classified as *Homo sapiens neanderthalensis,* and the title of the Cro-Magnons is *Homo sapiens sapiens.*

The Cro-Magnons, a representative of our own species, *Homo sapiens,* can be traced back about 35,000 years. Their remains have been found in

Figure 16.10 **Cro-Magnons,** representatives of our own species, *Homo sapiens,* can be traced back about 35,000 years ago in Europe. (Painting by Maurice Wilson; by permission of the Trustees of the British Museum—Natural History.)

many sites in western and central Europe. The Cro-Magnons showed individual differences, just as humans today exhibit individual variation. One of the notable variants is Combe-Capelle from France (fig. 16.10). Little is known of modern people in other continents during the time that Cro-Magnon flourished in Europe. We do not know the birthplace of modern *Homo sapiens.* Modern humans may have been cradled in Asia or Africa.

Pleistocene Overkill

The Cro-Magnons were accomplished hunters, so much so that they have been held responsible for the high rate of destruction and extinction of the mammalian fauna—such as the giant sloth, mammoths, saber-toothed cat, and giant ox—in the upper Pleistocene. The large-scale annihilation of many game mammals has been called *Pleistocene overkill* by the evolutionist Paul S. Martin. Martin has noted that more than a third of the genera of large mammals met extinction in Africa about 50,000 years ago, and that nearly all of the larger game genera in North America perished 12,000 to 15,000 years ago. The latter event coincided with the migrations of early Mongoloid bands into North America via the Bering Strait. So devastating were the hunting activities of humans in North America that only a few forms of larger mammals survived, notably the bison and pronghorn antelopes. Such drastic reductions in game animals throughout the world may have been one of the precipitating factors in the transition some 10,000 years ago from a hunting economy to an agricultural economy.

Problems of Present-day Humans

The uniqueness of the human species resides not only in its biological attributes but also in its capacity for cultural adaptation. A cultural heritage is not shared by any other animal on this earth. Other forms of life have evolved and have spread by adaptive radiation into different available ecological situations. Most species of organisms have become highly specialized, and many have passed unavoidably into oblivion with the onset of radical environmental changes to which the species has been unable to adjust or adapt. Indeed, a cardinal theme of evolution is that most species of organisms eventually decline and pass out of existence. There is accordingly no natural law or force that assures the future survival of any species, including the human species.

In a large sense, the human species is averting, or at least forestalling, extinction by cultural adaptation. As a consequence of our higher mental faculties, we have been able, through cultural endowments, to supplement and transcend our purely biological inheritance. Human culture has emerged out of our exercise of reason and our ability to communicate rational thoughts. The child is born uncivilized and acquires customs, beliefs, and values by instruction and imitation. Each new generation can draw on the rich store of past accumulated knowledge and ideas.

Through a capacity to absorb, transmit, and modify the body of learned tradition known as culture, we have been able to reach beyond ourselves. In contrast to other, humbler animals, we can imagine and plan. We are the only animal who can write a book, and we are the only animal who can assemble in conferences to discuss our goals and achievements. Each new success only serves to bring to light new areas of difficulty. Considering these facts, it seems inevitable that we are uncertain about our capacities and limitations, and hence set our goals either too high or too low. Our ambivalence concerning our goals forever plagues us.

Selected Readings

Birdsell, J. B. 1974. *Human evidence.* Chicago: Rand McNally & Co.

Campbell, B. G. 1966. *Human evolution.* Chicago: Aldine Publishing Co.

Clark, W. E. LeGros. 1964. *The fossil evidence for human evolution.* Chicago: University of Chicago Press.

Eimerl, S.; DeVore, I.; and editors of Time-Life Books. 1965. *The primates.* New York: Time-Life Books.

Goodman, M., and Tashian, R. E. (eds.) 1976. *Molecular anthropology: evolving information molecules in the ascent of the primates.* New York: Plenum Press.

Howell, F. C., and editors of Time-Life Books. 1965. *Early man.* New York: Time-Life Books.

Howells, W. W. 1959. *Mankind in the making.* Garden City, N.Y.: Doubleday & Co.

———. 1960. The distribution of man. *Scientific American,* September, pp. 112–27.

———. 1966. *Homo erectus. Scientific American,* November, pp. 46–53.

King, M-C, and Wilson, A. C. 1975. Evolution at two levels in humans and chimpanzees. *Science* 188:107–16.

Johanson, D. C., and White, T. D. 1979. A systematic assessment of early african hominids. *Science* 203:321–30.

Leakey, R., and Lewin, R. 1977. *Origins.* New York: E. P. Dutton & Co.

Lee, R. B., and DeVore, I. 1968. *Man the hunter.* Chicago: Aldine Publishing Co.

Lowenstein, J. M. and Zihlman, A. C. 1984. Human evolution and molecular biology. *Perspectives in Biology and Medicine* 27:611–22.

Medawar, P. B. 1961. *The future of man.* New York: The New American Library.

Montagu, A. 1962. *Man: his first million years.* New York: The New American Library.

Morris, D. 1968. *The naked ape.* New York: McGraw-Hill Book Co.

———. 1969. *The human zoo.* New York: Dell Publishing Co.

Napier, J. 1967. The antiquity of human walking. *Scientific American,* April, pp. 56–66.

Pearson, R. 1974. *Introduction to anthropology.* New York: Holt, Rinehart & Winston.

Pfeiffer, J. E. 1969. *The emergence of man.* New York: Harper & Row.

Pilbeam, D. R. 1972. *The ascent of man.* New York: Macmillan Co.

Potts, R. 1984. Home bases and early hominids. *American Scientist* 72:338–47.

Shepard, P. 1973. *The tender carnivore and the sacred game.* New York: Charles Scribner's Sons.

Simons, E. L. 1964. The early relatives of man. *Scientific American,* July, pp. 50–62.

Trinkhaus, E., and Howells, W. W. 1979. The neanderthals. *Scientific American,* December, pp. 118–33.

Walker, A., and Leakey, R. E. F., 1978. The hominids of East Turkana. *Scientific American,* August, pp. 54–66.

Washburn, S. L. 1960. Tools and human evolution. *Scientific American,* September, pp. 62–75.

————. 1978. The evolution of man. *Scientific American,* September, pp. 194–208.

Young, L. B., ed. 1970. *Evolution of man.* New York: Oxford University Press.

Zihlman, A., and Lowenstein, J. M. 1979. False start of the human parade. *Natural History* 88:86–91.

Zihlman, A., and Tanner, N. 1978. Gathering and the hominid adaptation. In *Female hierarchies,* ed. L. Tiger and H. Fowler, pp. 163–93. Chicago: Beresford Book Service.

17
Microevolution, Macroevolution, and Punctuated Equilibria

Evolutionary changes in populations are ordinarily visualized as gradual, built upon many small genetic variations that arise and are passed on from generation to generation. The shifting gene frequencies in local populations may be thought of as *microevolution*. The progressive replacement of light-colored moths by dark moths in industrial regions of England exemplifies the microevolutionary processes. Most population geneticists subscribe to the view that the same microevolutionary processes have been involved in the major transformations of organisms over long spans of geologic time (*macroevolution*). The traditional outlook is that small variations gradually accumulate in evolving lineages over periods of millions of years. If we were to recover a complete set of fossil specimens of a lineage, we would expect to find a graded series of forms changing continuously from the antecedent species to the descendant species. The temporally intermediate forms would be intermediate morphologically. Under this perspective, our inability to find transitional fossil forms between the ancestral and descendant populations would represent merely the imperfect nature of the fossil record.

For several years, Stephen Jay Gould of Harvard University and Niles Eldredge of the American Museum of Natural History have questioned the conventional view that evolutionary changes in the distant past are principally the outcome of the gradual accumulation of slight inherited variations. They advocate that most evolutionary changes have consisted of rapid bursts of speciation alternating with long periods in which the individual species remain virtually unmodified. Gould and Eldredge maintain that most lineages display such limited morphological changes for long intervals of geologic time as to remain in stasis, or in "equilibria." Conspicuous or prominent evolutionary changes are concentrated in those brief periods ("punctuations") when the lineages actually split or branch. This is the now celebrated hypothesis of *punctuated equilibria.*

Rapid Evolution in Animal Species

In the discussion of geographic speciation in chapter 9, there was scarcely any hint that a new animal species could arise abruptly. Species formation was held to occur by the gradual genetic divergence of geographically separated (allopatric) populations. In the 1970s, Hampton L. Carson of the University of Hawaii in Honolulu demonstrated that animal speciation can occur rapidly, within relatively few generations. This has been evidenced in populations of flies of the famliy Drosophilidae, commonly called the "picture-winged" flies. These extraordinary flies are indigenous to the high-altitude forests of the Hawaiian Islands (Archipelago) in the tropical region of the Pacific Ocean. The geologically very recent Hawaiian Islands, created by volcanic action, are the most isolated ocean islands in the world, at least 2,000 miles from any continent. The youngest and largest island in this archipelago is Hawaii, which has received much of its unique fauna from the adjacent island of Maui.

Chance has played the major role in the establishment of the interesting array of species of picture-winged flies on Hawaii. There are 26 species of pictured-winged flies inhabiting an island that is less than 700,000 years old. On the basis of analysis of chromosomal inversions, Carson infers that each species on Hawaii was descended from a few waif ancestors (indeed, possibly a single gravid female) that arrived fortuitously from Maui and other older islands. After each successful colonizing event, morphologically and genetically distinct species have been formed. An important aspect is that the migrant populations gave rise to new species rather than simply to new colonies of the parental species.

The process of almost instantaneous speciation seems to have occurred repeatedly in the proliferation of Hawaiian flies. Each small isolate of the ancestral population contained only a small fraction of the ancestral gene pool. From such a disrupted or reduced gene pool, natural selection rebuilt and reorganized a genetic system that contained unique elements and combinations. New genotypes arose that would not have appeared or would not have survived in the parental populations. Reproductive isolation most likely developed largely as an incidental by-product of the new gene complexes. It is striking that the evolutionary events that produced the varied species of picture-winged flies on Hawaii consumed less than 700,000 years.

The Founder Principle and Rapid Diversity

It may be recalled (chapter 8) that the Founder principle is a special case of Sewall Wright's genetic drift. The Founder principle is the establishment of a new population by a few migrant individuals who carry only a small fraction of the total genetic variability of the parental population. As first enunciated by the Harvard evolutionist Ernst Mayr in 1956, such small, peripherally isolated populations may undergo rapid evolutionary divergence to form new species. In fact, such rapidly evolving, locally isolated populations may have been the source of many evolutionary novelties in the distant past. The genetic restructuring of a Founder population is likely to be novel, since strong selection pressures are operating on an altered gene pool in an alien habitat. Morphological differentiation often occurs in conjunction with the evolution of reproductive barriers separating the colonizing population from the parent population. Mayr's scheme of the rapid emergence of a new species by the Founder principle provides one of the essential ingredients in the punctuational mode of evolution expounded by Gould and Eldredge.

Tenet of Punctuated Equilibria

A single line of descent, or lineage, may persist for long reaches of geologic time. As small changes accumulate over periods of millions of years within one lineage, the descendant populations may eventually be recognized as a species distinct from the antecedent populations. The persistent accumulation of small changes within a lineage has been termed *phyletic gradualism,* and the transformation of a lineage over time has been termed *anagenesis.* As depicted in figure 17.1, a new species (labelled "B") arises from the slow and steady transformation of a large antecedent population ("A"). If only transformation occurred, then life would cease as single lineages became extinct. Hence, a new species ("C" in fig. 17.1) can also arise by the splitting or branching of a lineage. The splitting of one phyletic lineage into two or more lines is termed *cladogenesis.* But, here again, the splitting is conceived of as proceeding slowly and gradually, with the two branching lineages progressively diverging, without significant reduction in population size. Thus, paleontologists tend to view lineage splitting in terms of gradual morphological divergence. Viewed as a slow process, lineage splitting becomes reduced to a special case of the phyletic model.

Gould and Eldredge maintain that phyletic gradualism is much too slow to produce the major events of evolution. In fact, the morphological changes in successive populations of a lineage are usually directionless and only minor temporal variations occur. The theory of punctuated equilibria advocates that the prominent episodes of evolution in the history of life are associated with the splitting of lineages, but the splitting is not seen as slow and steady cladogenesis. The new species arises through rapid evolution when a small local population becomes isolated at the margin of the geographic range of the parent species (fig. 17.2). Indeed, the successful branching of a small isolated population from the periphery of the parental range virtually assures the rapid origin of a new species. In geologic time, the branching is sudden—thousands of years or less, compared to the millions of years of longevity of the species itself.

Figure 17.1 According to phyletic gradualism, a lineage consists of a graded series of intermediate forms connecting ancestral and descendant organisms. Species A gradually transforms into species B, and may split into a slowly evolving species C.

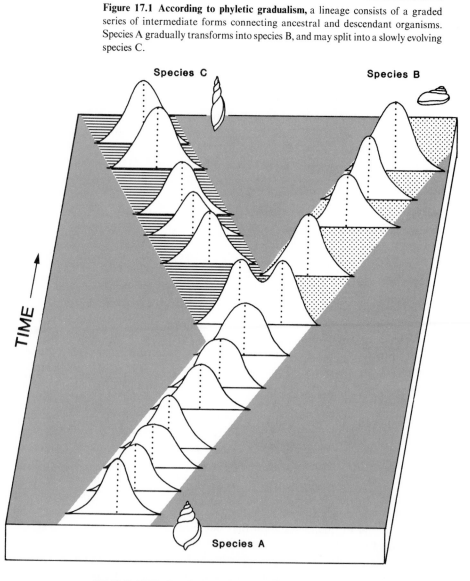

PHYLETIC GRADUALISM

Figure 17.2 The theory of punctuated equilibria proposes that a small isolate from the parental population (species A) evolves rapidly into a new entity (species B), which undergoes a long period of stasis during which little or no morphological change takes place. Species B does undergo minor structural modifications with time, but these modifications represent merely oscillations around the mean.

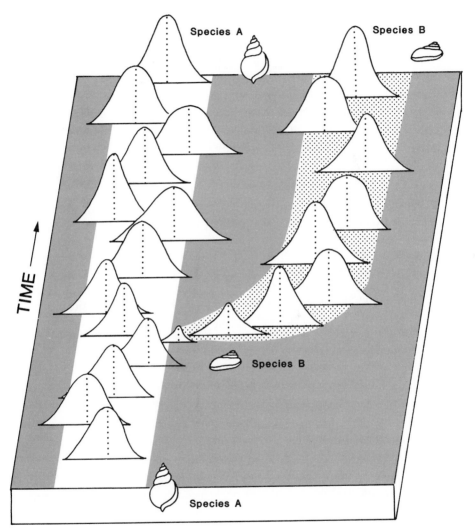

PUNCTUATED EQUILIBRIA

The punctuational change does not entail any unconventional evolutionary phenomenon. The sudden appearance of a species is fully consistent with Ernst Mayr's rapid selection in peripheral isolates. The morphological differentiation of geographical isolates occurs early and rapidly, especially when the population is small and adapting to the new atypical environment at the periphery of the range. Once established as a new species, the expectation is that little evolutionary change will occur in that species over geologic time. The long stasis of fossil species is disconcerting to evolutionists who endorse the slow phyletic transformation of species.

Given the thesis of punctuated equilibria, the fossil record is a faithful rendering of the evolutionary processes. If a species remains essentially static during its long lineage, then one would not anticipate finding a continuous series of transitional fossils. Moreover, if a descendant species arises by allopatric speciation in small, peripheral populations, then it would be rare to find evidence of preservation of a small population evolving in a relatively brief period of time. Thus, the so-called "gaps" in the fossil record may not be gaps at all. Gaps can not be expected to be filled with intermediate forms that do *not* exist.

Perhaps the best documented case of the punctuational mode of evolution is the study of Peter G. Williamson in 1981 of fossilized snails in the Lake Turkana region of Africa. The thick, undisturbed fossil beds contain millions of preserved shells representing at least 19 species of snails. Several lineages of Cenozoic snails were exceptionally stable for 3–5 million years. When morphological changes in shell shape did occur, they were concentrated in brief periods of 5,000–50,000 years. These rapidly evolved new populations then persisted virtually unmodified until they became extinct. Some authors have cogently commented that so-called sudden changes within 50,000 years might appear to be instantaneous to a paleontologist but would be almost an infinity to an experimental geneticist. Indeed, to many geneticists, an interval of 50,000 years (or 20 generations in snails) would be sufficient time for the morphological changes in the snail populations to accrue gradually rather than dramatically.

"Hopeful Monster"

There are genetic changes, known as *homeotic* mutations, which are responsible for such dramatic developmental effects as the transformation of one major organ into another. In the fruit fly, homeotic mutations can transform an antenna into a leg, or a balancer into a wing, or an eye into an antenna. Although fascinating for developmental studies, most evolutionists would not even contemplate that the origin of a new species could ever involve such drastic major morphological changes produced by single homeotic gene changes, or *macromutations*. However, in the 1940s, the prominent geneticist Richard Goldschmidt contended that homeotic mutants do illustrate a possible mode of origin of novel types of body structure, and hence could be responsible for the rapid emergence of a new species in the distant past. He conceded that the vast majority of macromutations would have disastrous consequences ("monsters"). An occasional macromutation, however, might conceivably immediately adapt the organism to a new way of life, a "hopeful monster" in Goldschmidt's terminology.

Goldschmidt was convinced that the gradual accumulation of small mutations was too slow a process to account for broad macroevolutionary trends. Goldschmidt favored the idea of a dramatic, abrupt transformation of a population by a favorable, instantaneous macromutation, which could take the form of a radical rearrangement of genetic material (a gross chromosomal change). The concept of punctuated equilibria, as formulated by Gould and Eldredge, does *not* require or imply fortuitous favorable macromutations. Rather, in the process of allopatric speciation in peripheral isolates, the incipient species develop their distinctive features rapidly. However, the importance of one aspect of Goldschmidt's views cannot be denied: structural chromosomal rearrangements may play a primary role in the initiation or achievement of reproductive isolation between two diverging populations.

Evolution of Horses

In reconstructing the phylogeny of an animal group, it has been the standard practice to show the emergence from a central, generalized stock of a large number of divergent branches or lineages. Not all branches persist; indeed, the general rule is that all but a few perish. The disappearance of many branches in the distant past might lead the observer today to the mistaken impression that the evolution of a particular group was not at all intricately forked. Thus, the evolution of horses could be erroneously depicted as an undeviating, straight-line progression from the small, terrier-sized *Hyracotherium* (formerly known as *Eohippus,* the "dawn horse") to the large modern horse, *Equus*. On the contrary, a wealth of evidence has revealed convincingly a pattern of many divergent lineages (fig. 17.3).

How does one interpret the divergent lineages? When new fossil discoveries revealed that a single, continuous lineage was implausible, proponents of phyletic gradualism simply postulated the existence of more than one line of gradual descent. We may dwell on one important lineage. The appearance of the one-toed condition was a landmark in horse history. The fleet-footed, one-toed *Pliohippus* emerged during the Pliocene epoch from the slow-footed, three-toed *Merychippus* (fig. 17.3). This major event is usually interpreted as a gradual change within a single lineage. In other words, a direct ancestral-descendant relationship between *Merychippus* and *Pliohippus* has been widely accepted. The paleontologist George Gaylord Simpson departed from the view of slow and steady change when he suggested that such a major transition might have involved the accumulation of small genetic changes in relatively rapid succession. This accelerated pace of phyletic gradualism was called "quantum evolution" by Simpson.

Figure 17.3 Phylogenetic tree of horses through time, with its many divergent branches. All branches died out, save one which eventually culminated in the modern group of horses, *Equus*. The history of horses dates back to early Cenozoic times, some 60 million years ago. (The Cenozoic *era* is divided into *periods,* which in turn are subdivided into *epochs.*)

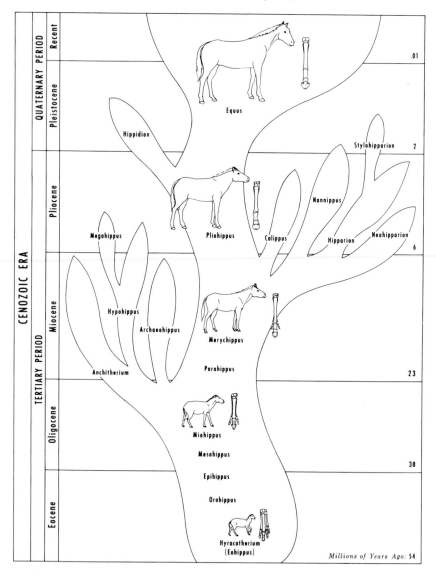

A geologically sudden appearance of *Pliohippus,* as prescribed by the model of punctuated equilibria, is also plausible. One might argue that *Merychippus* existed for a long period with little or no evolutionary change (stasis), and eventually suffered extinction. *Pliohippus* is not a linear descendant of *Merychippus,* but a dramatically new form that had earlier arisen as a small peripheral population of the parent species. It may be noticed (fig. 17.3) that not all offshoots of *Merychippus* evolved the progressive single-toed condition. Several lines of Pliocene horses, such as *Hipparion* and *Nannippus,* retained the conservative three-toed pattern. These conservative Pliocene horses were evolutionary blind alleys, and may well have represented "punctuations" that failed.

Modes in Human Evolution

In the past two decades, Africa has become increasingly the focus for investigations into human origins and differentiation. Several sites in Africa have yielded fossils that reveal different phases of the human story from the earliest stages. The differences of opinion that exist concerning the phylogeny of the Hominidae are vividly revealed by the ever-changing reconstructions of the human and evolutionary tree, almost on a yearly basis (see chapter 16). One of the most widely held views is that a gradual ancestral-descendant relationship existed for each successive hominid group. Thus, the australopithecines shaded imperceptibly into *Homo habilis,* who in turn graded slowly into *Homo erectus,* with the latter ultimately transforming into modern *Homo sapiens.* The theory of punctuated equilibria claims that the evolutionary changes in hominids occurred in rapid bursts, concentrated in speciation events (fig. 17.4).

Figure 17.4 The rectangular pattern of hominid phylogeny as expressed by the controvertible theory of punctuated equilibria.

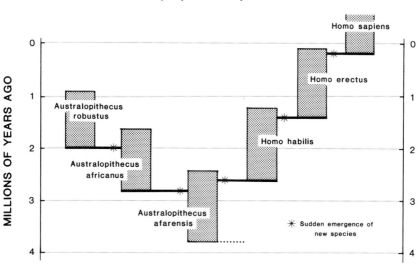

Such punctuated episodes were separated by long periods of stasis in the hominid lineages during which little or no morphological change took place. Figure 17.4 shows that this pattern of evolution appears "rectangular" rather than tree-like.

Evolution through punctuated equilibria is a reasonable, but controversial, hypothesis. It remains to be seen whether the fossil record for many organisms does, in fact, show that the punctuational mode plays a comparatively major role in the origin of evolutionary novelty.

Selected Reading

Carson, H. L. 1982. Evolution of Drosophila on the newer Hawaiian volcanoes. *Heredity* 48:3–25.

———. 1982. Speciation as a major reorganization of polygenic balances. In White, M. J. D., ed., *Mechanisms of speciation*. New York: Alan R. Liss, Inc., pp. 411–33.

———. 1983. Chromosomal sequences and interisland colonizations in Hawaiian Drosophila. *Genetics* 103:465–82.

Cronin, J. E.; Boaz, N. T.; Stringer, C. B.; and Rak, Y. 1981. Tempo and mode in hominid evolution. *Nature* 292:113–22.

Goldschmidt, R. 1940. *The material basis of evolution*. New Haven, Conn.: Yale University Press.

Gould, S. J. 1976. Ladders, bushes, and human evolution. *Natural History* 85(4):24–31.

———. 1977. Caring groups and selfish genes. *Natural History* 86(10):20–24.

———. 1977. Evolution's erratic pace. *Natural History* 86(5):12–16.

———. 1977. The return of hopeful monsters. *Natural History* 86(6):22–30.

———. 1980. Is a new and general theory of evolution emerging? *Paleobiology* 6:119–30.

———. 1982. Darwinism and the expansion of evolutionary theory. *Science* 216:380–87.

———. 1982. The meaning of punctuated equilibrium and its role in validating a hierarchical approach to macroevolution. In Milkman, R., ed. *Perspectives on evolution*. Sunderland: Sinauer Associates, pp. 83–104.

———, and Eldredge, N. 1977. Punctuated equilibria: the tempo and mode of evolution reconsidered. *Paleobiology* 3:115–51.

Herbert, W. 1983. Lucy's family problems. *Science News* 124:8–11.

Johanson, D. C., and White, T. D. 1979. A systematic assessment of early African hominids. *Science* 203:321–30.

Leakey, R. E. F. 1976. *Australopithecus, Homo erectus* and the single species hypothesis. *Nature* 261:572–76.

Lewin, R. 1980. Evolutionary theory under fire. *Science* 210:883–87.

Mayr, E. 1963. *Animal species and evolution*. Cambridge: Harvard University Press.

Schwartz, J. H. 1984. The evolutionary relationships of man and orang-utans. *Nature* 308:501–05.

Simpson, G. G. 1944. *Tempo and mode in evolution.* New York: Columbia University Press.

———. 1953. *The major features of evolution.* New York: Columbia University Press.

Smith, J. M. 1983. The genetics of stasis and punctuation. *Annual Review of Genetics* 17:11–25.

Stanley, S. M. 1979. *Macroevolution.* San Francisco: W. H. Freeman Co.

Stebbins, G. L. and Ayala, F. J. 1981. Is a new evolutionary synthesis necessary? *Science* 213:967–71.

Templeton, A. R. 1980. The theory of speciation via the founder principle. *Genetics* 94:1011–38.

———. 1981. Mechanisms of speciation—a population of genetic approach. *Annual Review of Ecology and Systematics* 12:23–48.

Turner, J. 1984. Why we need evolution by jerks. *New Scientist* 101:34–35.

Whitten, P., and Nickels, M. K. 1983. Our forebears' forebears. *The Sciences,* January-February, pp. 20–21.

Williamson, P. G. 1981. Palaeontological documentation of speciation in Cenzoic molluscs from Turkana Basin. *Nature* 293:437–43.

Wolpoff, M. H. 1968. "Telanthropus" and the single species hypothesis. *American Anthropologist* 70:477–93.

18
Natural Selection and Social Behavior

In his provocative book entitled *Sociobiology,* Edward O. Wilson of Harvard University delves imaginatively into the social interactions of all animals, including humans. The cardinal theme is that the social interplay of animals, no matter how complex, has evolved by natural selection. Accordingly, socio-biologists attempt to place social behavior on sound Darwinian principles. The doctrine of natural selection is thought of as central to the biological under-standing of sociality.

If the behavioral attributes of animals are the products of the same evolutionary forces that shape morphological and physiological traits, then social behavioral patterns should be adaptive. That is to say, behavioral dis-positions should be optimally designed to confer reproductive success upon the individual. There are, however, behavioral acts that are detrimental to the individual performing the act, but which promote the reproductive advantage of other members of the population. A female worker bee, for example, com-pletely forsakes procreation and labors ceaselessly to enhance the reproductive fitness of the queen. Is it possible to select for behaviors that are individually disadvantageous but beneficial to the species as a whole? This is one of the searching questions in sociobiology today.

Group Selection

In 1932, the English geneticist J. B. S. Haldane speculated on the possibility that a trait may be selected that confers an advantage for the group but is costly to the individual. He used the term *altruism* for such a trait, and defined an altruistic act as one that decreases the personal fitness of the organism performing the act, but is beneficial to the population as a whole. Thus, a prairie dog that emits a loud warning call (or alarm) when it spots a coyote improves the chances that his fellow conspecifics will survive. In protecting the group, the prairie dog (alarmer) attracts attention to itself and places itself in immediate danger of being captured by the predator. If the prairie dog were to remain silent, its presence to the coyote would not be betrayed. The warning act by the alarmer is essentially self-sacrificial; the term *altruism* may be equated with self-sacrificial or selfless. Haldane acknowledged the difficulty in explaining, in a mathematical model, the establishment of an altruistic trait that apparently diminishes one's own chances of survival and reproduction.

In 1962, the Scottish ecologist V. C. Wynne-Edwards evoked the concept of *group selection* to account for altruism as it relates to territorial behavior. Wynne-Edwards views territoriality as a method of population control. When a given population becomes excessive, many individuals cannot find territories and therefore cannot breed. Wynne-Edwards suggests that the territorial system has evolved by natural selection as part of a mechanism to stabilize the population density at a level that can be supported by the available food resources of the area. Wynne-Edwards' thesis assumes that natural selection operates for the benefit of the group as a whole. The implication is that many individuals are genetically or internally programmed to refrain from reproducing so as not to endanger the welfare or survival of the stock. In other words, natural selection has fostered both territorial "winners" and "losers" for the good of the species as a whole. Stated another way, a "loser" promotes the reproductive advantage of the "winner" at its own expense.

The curtailment of reproductive activity in "losers" is inconsistent with the notion of individual selection. An abridgment of reproductive behavior certainly does not benefit the individual. Natural selection characteristically operates on individuals (*not* groups) to augment (*not* decrease) individual reproductive capability. Since selection presumably operates solely to maximize the reproductive success of each individual, it is difficult to imagine how individuals can be selected to save the group at their own individual expense.

From a theoretical standpoint, can reproductive curtailment or restraint on the part of an individual evolve in a population so as to confer a reproductive advantage upon the group as a whole? Let us suppose that gene a_1 promotes reproductive capability and its allele, a_2, tends to curtail the reproductive capacity of an individual. Can the a_2 allele persist, or even spread, in a population when its effect is to impair the reproductive fitness of its possessor in the present and succeeding generations? The a_2 allele necessarily is selected against, since the possessors of the alternative allele, a_1, obviously leave more offspring than the possessors of the a_2 gene. Ultimately, the self-sacrificial a_2 allele will be replaced by the reproductively advantageous a_1 gene.

Kin Selection

Altruism apparently cannot evolve without violating the principle of individual selection, since the individuals displaying self-sacrificial behavior are less biologically fit than their selfish (or egoistical) colleagues. Yet altruistic behaviors are clearly evident, particularly among birds and mammals. A female bird behaves altruistically when she protects her brood against predation; a male baboon will emerge from the heart of a troop to attack a leopard that threatens the group; and a human will place his or her life in jeopardy to rescue a drowning person. In 1964, W. D. Hamilton of the London School of Economics proposed an alternative route for the evolution of altruism that is not founded on the indefensible premise of group selection.

Hamilton explained altruism as the outcome of a selective process called *kin selection*. The concept is based on the fact that close relatives share a high proportion of the same genes. Altruistic behavior can be favorably selected if the probability is high that the beneficiaries of the altruistic act also have the same genes as the self-sacrificial altruist. Under this view, kin selection is a special manifestation of *gene* selection. It is the gene coding for a particular behavior that is optimized or favorably selected. A given allele a_1 will spread if the behavior associated with this gene adds a greater number of a_1 alleles to the next generation than in the preceding generation. This can happen only if the reproductively successful individuals are close relatives of the altruist, thereby increasing the probability that they carry the same a_1 allele.

An instructive phenomenon is the special chirp, or warning call, in bird populations, by which one member alerts the flock to the danger of a predator. We may assume that the warning note is governed by an a_1 allele. A warning call originating as a signal from a parent to its offspring during the breeding season is selectively advantageous, because the degree of relatedness between the caller and recipient is high. Gene a_1 will persist in the population, since the alarm call by the parent (bearing gene a_1) increases the fitness of sufficient numbers of offspring also possessing a_1. By protecting the offspring, the parent invests in its own genetic representation in subsequent generations. In an evolutionary sense, the offspring that are protected by the parent are, in part, genetically the parent itself.

The lifetime of an organism is to be viewed as a strategy for perpetuating the organism's genes. An individual can transmit its genes directly through its own reproduction as well as by proxy through its close relatives who share genes by common descent. An individual is said to maximize an *inclusive fitness*—its own reproductive fitness plus the reproductive fitness of close relatives. The degree of genetic relatedness has a crucial bearing on the likelihood that two individuals will behave altruistically toward one another. The extent to which two individuals share genes by descent from a common ancestor is referred to as the degree of relatedness, or coefficient of relationship (designated r). In diploid species, the degree of relatedness of an individual to his or her full siblings is $1/2$; to half-siblings, $1/4$; to children, $1/2$; and to first cousins, $1/8$. Stated another way, a given individual shares 50 percent of his genes with a full brother or sister but only 12.5 percent of his genes with a first cousin. By this logic, grandparents and grandchildren are related by $1/4$.

Kin selection molds a form of altruism that is channeled to genetic relatives. An altruistic gene will be perpetuated if the reproductive benefit gained by the recipient of the altruistic act exceeds the cost in reproductive fitness suffered by the altruist. Figure 18.1 depicts kin selection with respect to two

Figure 18.1 Sibling altruism. The coefficient of relationships between two brothers is ½. Thus, the probability is 0.50 that a gene (or chromosome) present in one individual also occurs in his brother. For simplicity, only a pair of chromosomes is depicted. The individual's altruism to his brother is selectively advantageous because the altruistic act enables his brother to transmit greater numbers of replicas of the shared chromosome to the next generation than the individual himself might have transmitted.

Brothers Share 50% of Their Genes

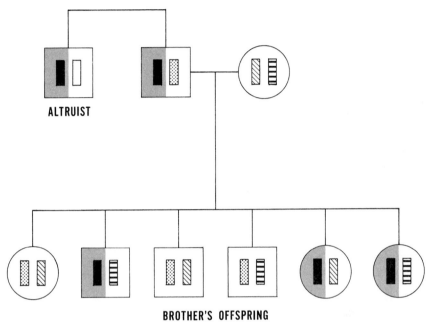

brothers. The altruist leaves no offspring, but his sacrificial act enables his brother to leave more offspring than he would have otherwise. The altruist's genes have been effectively removed from the population by his failure to reproduce, but the genes that he shares with his brother are more than restored by three of the offspring sired by his brother. The altruist has actually gained genetic representation in the next generation. Indeed, he has gained in *inclusive fitness,* even though he has lost fitness in the classical sense.

Kin selection provides an explanation for the baffling social behavior in ants, bees, and wasps. Curiously, whole castes of sterile females devote their entire existence to the welfare of the queen. Females are diploid individuals that develop from fertilized eggs with maternal and paternal sets of chromosomes. Males are the haploid products of unfertilized eggs and possess only the maternal set of chromosomes (fig. 18.2). The startling outcome is that the sibling daughters of a queen are more closely related to each other than they would be to any of their own daughters!

Figure 18.2 Degree of genetic relationship of a female bee to her offspring *(top)* and the genetic relatedness of sisters to sisters *(lower left)* and sisters to brothers *(lower right)*. Females develop from fertilized eggs and have two sets of chromosomes (diploid state), whereas males develop from unfertilized eggs and have only one set of chromosomes (haploid state). For simplicity, the haploid set is represented by only two chromosomes. Sisters are more closely related to one another than they would be to their own offspring. The coefficient of relationship between mother and daughter is ½, but the coefficient of relationship between two sisters averages ¾. Sisters have relatively few genes in common with their brothers, the coefficient of relationship averaging ¼.

MOTHER-OFFSPRING RELATEDNESS

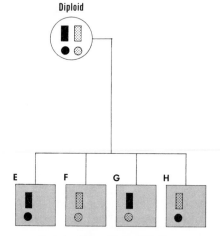

Each Daughter Shares 50% of Her Genes
With Her Mother

Each Son Shares 100% of His Genes
With His Mother

SISTER-SISTER RELATEDNESS

SISTER-BROTHER RELATEDNESS

Specifically, female workers share three-fourths of their genes with their sisters ($r = 3/4$). This is the case because sisters share the same set of paternal genes and share additionally, on the average, one-half of the maternal genes (fig. 16.2). Thus, sisters are related by the average of 1 (for paternal genes) and 1/2 (for maternal genes), or 3/4. If a daughter were to produce its own offspring, that daughter would share only one-half of its genes with any of its offspring. Accordingly, a female worker actually contributes more to her Darwinian fitness by assisting her mother in raising offspring (3/4 relationship) instead of rearing her own offspring (1/2 relationship). Natural selection has fostered an unusual form of cooperative behavior among sisters.

The male bees, or drones, show a pronounced lack of concern over the welfare of their sisters. This is not surprising, since drones share only ¼ of their genes with their sisters. Since a drone possesses only maternal genes, a sister cannot share any paternal genes with her brother (fig. 18.2). The total genetic relationship of a sister to her brother is the average of zero (for paternal genes) and ½ (for maternal genes), or ¼. The drone is notoriously lazy ("lazy as a drone"), since his inclination is to produce daughters (who share all of his genes) rather than help his sisters. In turn, sisters invest more energy in raising sisters than brothers and characteristically drive their brothers from the hive in the early summer. The selfishness of the drones and the diligence of the female workers are predictable aspects of kin selection.

Kin Selection in Humans

A variety of altruistic behavioral dispositions in the human species—the sharing of food, the sharing of implements, and the caring of the sick—probably evolved by kin selection in the early hominid hunting bands. These bands consisted largely of tightly knit groups of close relatives. In fact, in primitive cohesive societies, social behavior appears to have been dominated by kin selection.

A striking phenomenon in some human societies is the strong parental attitude exhibited by a brother to his sister's children, far out of proportion to his degree of relatedness to his nephews and nieces. However, as predicted by kin selection, a brother's sense of responsibility to his sister's children increases as confidence in paternity of his own children diminishes. A brother is always genetically related to his sister's children, whereas he may be totally unrelated to his wife's children if the probability of paternity is low. Apparently, the certainty of a genetic relationship between brother and nephew (or niece) outweighs the dubious or tenuous relationship between father and alleged son (or daughter). In essence, a high probability of paternity is a necessary antecedent for extended parental investment by the male.

It has been suggested that homosexual behavior is a product of kin selection. Although the genetic factors for homosexuality lower reproductive fitness, the production of offspring occurs by proxy through the homosexual's immediate kin. In this vein, homosexuals may be compared to sterile worker bees who help raise their close genetic relatives. The behavior of the homosexual increases the fitness of genetic relatives more than it decreases his or her own fitness. This view is consistent with kin selection, but there are as yet no supporting or confirmatory data.

Kin selection may have shaped the phenomenon of menopause in the female. The middle-aged woman enhances her fitness by caring for her children's children. As the menopausal woman loses her capacity to have her own children, she regains inclusive fitness by devoting her efforts to the rearing of her grandchildren. It is interesting that the duration of fecundity is greater in women of economically advanced countries. Menarche occurs earlier and menopause is deferred. With improved nutrition and better personal hygiene, the female increases the duration in which she can invest in her own children.

The Selfish Gene

The alarm calls of birds are often cited as an excellent example of kin selection. It is generally the case that a close kin is sufficiently near the caller to benefit from the warning call. However, some authors contend that the habit of warning cries evolved by conventional individual selection rather than by kin selection. A warning call may *not* serve to draw the attention of the predator to the alarmer. The warning call may *not* expose the caller to greater danger. Rather, the call note is intended to mislead the predator as to the position of the prey.

Under this view, the alarmer utters a call that the predator cannot easily locate. Since the predator cannot readily detect the location of the call, it would appear that the alarmer issues the warning note to protect itself and not its colleagues (conspecifics). Suppose a flock of birds is feeding in a meadow, and a hawk flies past in the distance. The hawk has not seen the flock, but there is the danger that he will soon spot the flock. The noisy rummaging activity of members of the flock could attract the hawk's attention. The alarmer who first detects the hawk issues a warning call to its companions, but does so only to curtail the noise of the flock. Accordingly, the caller reduces the chance that the flock will inadvertently summon the hawk into his own vicinity. In large measure, the caller is not acting *altruistically* but rather *selfishly*.

The same situation may be viewed in another selfish vein. Suppose the caller fails to signal, quietly taking flight by itself without warning its unsuspecting conspecifics. In this event, the hawk might be easily attracted to a single bird flying off. The issuance of an alarm would be better than no alarm at all. An alarmer that issues a call curtails his risks by flying up into the tree—and simultaneously arouses the other members to fly with him into the tree. In corralling the conspecifics, the caller ensures protective cover for himself in flight. The conspecifics benefit from the caller's act, but in increasing his own safety, the caller benefits even more!

The portrait of a biological individual that emerges from these examples is one of self-serving opportunism. It seems that an individual is programmed to care about itself. It may be, as several investigators contend, that acts of apparent altruism are actually selfishness in disguise.

Reciprocal Altruism

When a self-sacrificial parent saves his own child from some potentially tragic event, the parent acts to protect his share of genes invested in the child. If, however, an individual saves the life of an unrelated party or nonrelative, kin selection is ruled out. In this instance, the individual may have behaved altruistically with the expectation that the beneficent behavior will be reciprocated by the stranger at some future occasion. A person who saves another person from drowning, for example, hopes in turn for help when he or she is in danger. This is the concept of *reciprocal altruism*, first proposed in 1971 by Robert L. Trivers of Harvard University. An altruist who places himself in danger for a biologically unrelated person incurs a reduction in fitness because of the energy consumed and the risks involved. However, a future reciprocal act by the recipient is likely to bring returns that are equal to, or greater than, the altruist's original expenditure.

There is definite survival value to both donor and recipient when altruistic acts are mutually exchanged. Each participant benefits by increasing his or her fitness. If, however, a recipient fails to reciprocate when the situation arises, that recipient would no longer have the benefits of future altruistic gestures to him. The nonreciprocator would then be at a selective disadvantage, since subsequent adverse effects on his life would not be overcome by altruistic acts that might be lifesaving. The ultimate effect of nonreciprocation would be the restriction of altruism to faithful fellow altruists, with the consequence that genes for reciprocal altruism would be perpetuated through the generations. In fact, the chances of selecting for altruistic genes are improved as more altruistic acts occur in the lifetime of the altruist. By the continual exchange of beneficent acts, altruists accrue more fitness in the long run than the nonreciprocators. Overall, altruists gain fitness, rather than lose fitness, from selfless acts.

"Stepmothering" may be considered a form of reciprocal altruism. When the biological mother dies or leaves her children, the male may remarry, and the stepmother assumes the care of the offspring. The basis of the reciprocity is best seen in the behavior of western gulls off the coast of California. As reported by Raymond Pierotti, an unmated gull, by raising the offspring of a previous mate, establishes a pair-bond relationship with the male and guarantees herself both a mate and a breeding site for subsequent years. Thus, in a population of gulls where there is a surplus of breeding-age females, the acquisition of a breeding territory complete with a mate is a compelling incentive for a female gull to behave altruistically by rearing another female's offspring on one occasion. By so doing, she increases her personal reproductive fitness for future years with only a relatively small initial investment. Such behavior is fostered by natural selection, since the final effect for the step-mother is an increase in her own genetic material in succeeding generations.

Parental Reproductive Strategies

In the placental mammals, the young develop slowly and the prolonged period of caring for the young is almost exclusively restricted to the female. The males of most mammalian species do not establish even a semblance of a permanent relationship with the pregnant and nursing female. In its extreme form, such as in cats and bears, the female actually ejects the male from any contact with the young. It is as if the mammalian male is biologically super-fluous after insemination. Apart from insemination, most mammalian females have little use for the males—except possibly to protect them from other males. It is only exceptionally, as in wolves and humans, that the male plays an important role in the protection and care of the mother and child.

Each sex, without conscious intent, strives to attain maximal reproductive success. The reproductive strategies of males and females are clearly different. Each sex may be viewed as an investor in which the capital invested is in terms of reproductive effort. The female undoubtedly invests more per off-spring than the male. The overproduced, lightweight sperm are inexpensive compared with the thriftily produced, energy-rich eggs. The prolonged internal gestation and the protracted period of maternal care of the newborn place rigorous demands of energy and time on the female. Accordingly, the female has a much greater stake in any one reproductive act. Since a reproductive mistake is much more severe for the female, she is more discriminating than the male in the choice of a mate.

A male's basic reproductive strategy is to achieve as many fertilizations as possible. Males are less selective in the choice of acceptable sexual partners and more aggressive in excluding other males from the opportunity to mate. Females are choosier and carefully evaluate their options. Males almost uni-versally act competitively as "sexual advertisers," and females act cautiously as "comparison shoppers." The task of critical discrimination of a sexual partner and the avoidance of an unproductive pregnancy resides almost wholly with the female.

The female selects that male as a sexual partner whose appearance and behavior signifies that he will transmit a superior set of genes to the offspring. Mammals are generally polygynous—that is, a single male mates with more than one female. As previously mentioned, most mammalian males do not establish any durable association with the female beyond the sexual act. If a reproductively fit mammalian male has little to offer except his superior genes, then it is to the female's advantage to mate with that suitable male, no matter how often he may have already mated. In polygynous mammalian species, the fittest males (at least those that are behaviorally the fittest) have the greatest reproductive success. For example, in breeding colonies of elephant seals, the males clash for dominance status, which is associated with mating rights. Less than one-third of the males in residence during a season copulate, and as few as five dominant males account for 50 percent of the matings in a given season. Clearly, females who share an oft-mated male of unusual competitive ability are attracted to a "winner" with "proven" genetic qualities.

The situation is different in birds, in which monogamy is the general rule. One male and one female form a breeding pair and remain together, at least for the rearing of one brood. In some species of birds, the same pair-bonding may persist for several successive broods (as in songbirds) or even for a lifetime (as in geese and swans). In these situations, the male's contributions to the female and her offspring are much more than a mere complement of genes. The males of many species of birds defend a territory, provide protection against predators, assist in the building of nests, incubate the eggs, and furnish nourishment for the female and her young. Thus, where the male provides numerous indispensable services, it is to the female's advantage to maintain pair-bonding. Stated another way, we can expect enduring pair-bonding only where the male plays a substantial paternal role.

The tendency toward polygyny in human males apparently has been dampened by strong selective pressures. The human male engages in an elaborate courtship ritual, which is in itself a substantial investment for sustained pair-bonding. Robert Trivers has suggested that the prolonged male courtship serves as insurance for the male that he is not establishing a bond with an already inseminated female. The greater the male investment, the greater the importance of paternity knowledge. To assure accurate identification of his own offspring, the male jealously guards against "cuckoldry"—that is, against the possibility that he might invest in offspring that are not his own. Human males universally treat their mates as possessions and sequester them from the adulterous interests of other males. It is only when the female remains faithful that the male can overcome his ever-present uncertainty about paternity. Nevertheless, the other side of the male's strategy is his inclination to cuckold other males. There have evolved, however, female counterstrategies to improve the female's ability to hold her mate's attention and care.

With the notable exception of humans, the female of mammalian species becomes sexually aroused only at certain seasons. At specific times the female comes into heat, or *estrus*, and only during these restricted periods is the female receptive to the male. The onset of heat is typically synchronized with ovulation, or the release of the ripe egg. In simultaneously advertising sexual receptivity and ovulation, the mammalian female provokes intense competition among the dominant males and increases her likelihood of securing a male of exceptional genetic constitution. The females are monopolized by the males only at ovulation.

In the human female, there are generally no outward or conspicuous signs of ovulation. Human females may be sexually receptive during any part of the ovulatory cycle. Several authors have pointed out that the loss of the phenomenon of estrus in human evolution was an event of great importance. The concealment of ovulation by the female serves to maximize her mate's confidence of paternity, in that potentially competing males are not flauntingly apprised of the ovulation event. Stated another way, if the human female were to glaringly advertise ovulation, it might cost her the protection and parental care of her mate by attracting competing males who threaten her mate's confidence of paternity. Moreover, her potentially philandering mate is restrained in seeking copulations with other fertile females who, like her, conceal the event of ovulation.

When the pair-bonding relationship fails or is dissolved, the female typically experiences difficulty in finding a substitute male to aid in the rearing of the offspring. We would expect strong selection pressure against "stepfathering," since the substitute father is called upon to invest in some other male's genes. The literature is replete with examples in varied mammals in which a male eliminates the offspring of a previous mating and replaces them with his own offspring. When the male lions of a new group depose the resident males of a pride, the new regime often kills the cubs of the displaced males. In another instance, a pregnant female mouse responds to the odor of a novel male by aborting her fetus and returning to estrus in preparation for propagating the genes of the new male. Among the Yanomama Indians of Venezuela, a man will order his newly acquired wife to kill her infants of a previous marriage. The biological urge to invest only in one's own offspring is evidently powerful and pervasive.

Parent-Offspring Conflict

Robert Trivers defines parental investment as the amount of care and assistance rendered by the parent that increases an offspring's chance of future reproductive success at the cost of limiting the parent's capacity to invest in other offspring. Each offspring is of equal value to a parent, since 50 percent of the parent's genes are distributed in each offspring. But each offspring is completely related to itself ($r = 1$), and related by one-half to each of its siblings ($r = 1/2$). Each offspring, then, would expect to receive twice as much parental care for itself as for each of its siblings. Thus, it is inevitable that conflicts will arise between parent and offspring.

In particular, a mother will normally apportion her investment among her offspring to maximize the number that survive to reproductive maturity. She must necessarily limit the amount of care that she can give to any one child. On the other hand, if the child is to maximize its own chances, the child will attempt to extract more than its share of parental care. The parental investment in an offspring eventually reaches a point of diminishing returns. That point is reached when the mother's investment in a given child costs her more than she gains and where it costs the child more than the child gains. Once the point of diminishing returns is met, the parent-offspring bond should be severed.

Implications for Society

Some writers declare that any behavioral act that increases biological fitness (individual or inclusive) is selfish and exploitative. The psychologist Donald Campbell of Northwestern University decries that man's basic behavioral tendencies are self-centered. Campbell advocates that society must counter the individual selfish tendencies that are promulgated by natural selection and, through learning and socialization, promote positive or beneficent expressions of social interplay. The implication is that our genetically based social tendencies are "bad" and only culturally imposed restraints are "good." Campbell presupposes that a social system works best by placing restraints on individual biological fitness. He also presumes that biological evolution and cultural evolution are sharply separated from one another. He thus revives the old, untenable nature-nurture dichotomy. This dichotomy is a trap of extremes. The extreme pessimistic view is that human behavior is solely influenced by genes, and any attempts to change behavior by learning are futile. The other extreme is the misguided optimism that humans are so infinitely plastic in social behavior that they can be molded easily to ideal social and economic systems.

In reality, the complex social behavior pattern of the human species is the outcome of the complementary interaction of biological evolution and cultural evolution. The key question is not whether there is a genetic foundation of social behavior in humans, but rather how firm and constraining is the foundation. The greater our knowledge and understanding of the degree of genetic determinism, the greater our capacity to modify behavior by cultural changes.

Richard D. Alexander of the University of Michigan asserts that we need not summarily suppress our biological inclinations. Human beings *do* behave in such a way as to maximize their genetic representation in future generations. Indeed, Alexander suggests that human culture is largely the manifestation of the reproductive strivings of the individual members of societies. Individuals, in maximizing their inclusive fitness, favor more closely related kin over more distantly related individuals. An individual's inclusive fitness is not generously served by aiding an unrelated person or stranger. An individual might aid a stranger on the expectation that such a person would ultimately, at some future date, reciprocate the aid in a similar circumstance.

The foregoing pattern of behavior is exemplified by a familiar example in our daily lives. We *felicitously give* Christmas presents to our close relatives. We *occasionally make a gift* to our more distant relatives. We *charitably exchange* Christmas cards with our acquaintances. Finally, we *unsparingly remove* from our list of card exchanges those acquaintances who fail to reciprocate.

Selected Readings

Alcock, J. 1975. *Animal behavior: an evolutionary approach.* Sunderland, Mass.: Sinauer Associated.

Alexander, R. D. 1974. The evolution of social behavior. *Annual Review of Ecology and Systematics* 5:325–83.

———. 1975. The search for a general theory of behavior. *Behavioral Science* 20:77–100.

Barish, D. P. 1977. *Sociobiology and behavior.* New York: Elsevier.

———. 1979. *The whisperings within: evolution and the origin of human nature.* New York: Harper & Row.

Campbell, B., ed. 1972. *Sexual selection and the descent of man 1871–1971.* Chicago: Aldine Publishing Company.

Campbell, D. T. 1975. On the conflicts between biological and social evolution and between psychology and moral tradition. *American Psychologist* 30:1103–26.

Chagnon, N. A., and Irons, W., eds. 1979. *Evolutionary biology and human social organization.* North Scituate, Mass.: Duxbury Press.

Daly, M., and Wilson, M. 1978. *Sex, evolution, and behavior.* North Scituate, Mass.: Duxbury Press.

Dawkins, R. 1976. *The selfish gene.* London: Oxford University Press.

Emlen, S. T., and Orling, L. W. 1977. Ecology, sexual selection, and the evolution of mating systems. *Science* 197:215–23.

Gregory, M. S.; Silvers, A.; and Sutch, D., eds. 1978. *Sociobiology and human nature.* San Francisco: Jossey-Bass Publ.

Hamilton, W. D. 1964. The genetical theory of social behavior. I and II. *Journal of Theoretical Biology* 7:1–52.

Hardin, G. 1977. *The limits of altruism.* Bloomington, Ind.: Indiana University Press.

Hinde, R. A. 1974. *Biological bases of human social behaviour.* New York: McGraw-Hill Book Co.

Hrdy, S. B. 1977. *The langurs of Abu.* Cambridge, Mass.: Harvard University Press.

Lorenz, K. 1966. *On aggression.* New York: Harcourt Brace Jovanovich.

Maynard-Smith, J. 1964. Group selection and kin selection. *Nature* 201:1145–47.

———.1978. The evolution of behavior. *Scientific American,* September, pp. 176–93.

Orians, G. N. 1969. On the evolution of mating systems in birds and mammals. *American Naturalist* 103:589–603.

Ruse, M. 1979. *Sociobiology: sense or nonsense?* Boston: D. Reidel Publ. Co.

Sahlins, M. 1977. *The use and abuse of biology: an anthropological critique of sociobiology.* Ann Arbor, Mich.: The University of Michigan Press.

Tiger, L., and Fox, R. 1972. *The imperial animal*. New York: Dell.

Trivers, R. I. 1971. The evolution of reciprocal altruism. *Quarterly Review of Biology* 46:35–57.

———. 1974. Parent-offspring conflict. *American Zoologist* 14:249–64.

Trivers, R. L., and Willard, D. E. 1973. Natural selection of parental ability to vary the sex ratio of offspring. *Science* 179:90–92.

Wilson, E. O. 1975. *Sociobiology: the new synthesis*. Cambridge, Mass.: Harvard University Press.

———. 1978. *On human nature*. New York: Bantam Books.

Wynne-Edwards, V. C. 1964. Population control in animals. *Scientific American*, August, pp. 68–74.

———. 1972. *Animal dispersion in relation to social behaviour*. New York: Macmillan Co., Hafner Press.

Index